ROUTLEDGE LIBRARY
ETHICS

Volume 12

ROLES AND VALUES

ROLES AND VALUES

An Introduction to Social Ethics

ROBERT (R. S.) DOWNIE

LONDON AND NEW YORK

First published in 1971 by Methuen & Co Ltd

This edition first published in 2021
by Routledge
2 Park Square, Milton Park, Abingdon, Oxon OX14 4RN

and by Routledge
52 Vanderbilt Avenue, New York, NY 10017

Routledge is an imprint of the Taylor & Francis Group, an informa business

British Library Cataloguing in Publication Data
A catalogue record for this book is available from the British Library

ISBN: 978-0-367-85624-3 (Set)
ISBN: 978-1-00-305260-9 (Set) (ebk)
ISBN: 978-0-367-90034-2 (Volume 12) (hbk)
ISBN: 978-1-00-302199-5 (Volume 12) (ebk)

Roles and Values

An Introduction to Social Ethics

R. S. Downie

METHUEN & CO LTD
11 NEW FETTER LANE, LONDON E.C.4

First published in 1971 by
Methuen & Co Ltd
11 New Fetter Lane, London E.C.4

Printed in Great Britain by
Richard Clay (The Chaucer Press) Ltd
Bungay, Suffolk

SBN 416 14910 3 hardback
SBN 416 14920 0 paperback

Distributed in the U.S.A.
by Barnes & Noble, Inc.

To Alison and Catherine

Contents

Preface

The aim of this book is to provide a philosophical investigation of the area in which the demands of social and political institutions impinge on individual values and responsibility. The concept of a social role is helpful in such an investigation because it enables the idea of institutional rights and duties to be combined with those of individual responsibility and values. My hope is that this approach to the problems of social ethics (by which is meant a mixture of social, political, and moral philosophy) will be of interest to students of the social sciences and of philosophy, and to all who have experienced the tension between individual values and institutional demands.

Thanks are due to two colleagues. The general philosophical viewpoint which forms the background to this essay was developed jointly with Miss Elizabeth Telfer in a previous work entitled *Respect for Persons*. Miss Telfer also discussed the details of this book and made many helpful suggestions. Miss Eileen Loudfoot provided criticisms of the theory of roles expressed in Chapter VI and drew my attention to many stylistic obscurities in other places, as well as assisting with the index and proofs.

R. S. DOWNIE

The University of Glasgow, 1970

1

The Nature of Social Ethics

1. SOCIAL ETHICS

What is 'social ethics'? There are a number of puzzles about the term, but let us begin by considering the meanings of the term 'ethics'.

In the first place, 'ethics' can be used simply as a synonym for 'morality'. It is indeed the mark of one kind of superior person to use the word 'ethics' instead of the word 'morality'. In the second place, 'ethics' can be used to mean 'morality' but with the added implication that, for once, it is not sexual morality which is under discussion. The appearance of this use reflects the fact that the terms 'morality' and 'immorality' have come to be increasingly overlaid with sexual significance. Thus, commenting on a political scandal, someone wrote to *The Times* that it was not the lying which he deplored so much as the immorality! Presumably he would call the lying simply unethical. Thirdly, 'ethics' is used to refer to codes of conduct which consist partly of ordinary moral rules, partly of rules of etiquette, and partly of rules of professional conduct. It is this sense which is in use when we speak of 'medical ethics' or 'the ethics of journalism'. Fourthly, 'ethics' is used to refer to the philosophical study of morality, which is also known as 'moral philosophy'. The word is used in this sense in titles such as *The Principles of Ethics*, and in the expression 'social ethics'. This book, then, is an introduction to the philosophical study of social morality, or at least, as we shall see, of one aspect of social morality.

But what is meant by a 'philosophical study'? A great deal can be said about this, but it may be that the best way of understanding the nature of philosophy is to engage in it. Hence, let us simply accept the provisional description of a philosophical investigation as one which attempts to analyse basic concepts and principles with the aim of clarifying them and relating them to one another. For example, a philosophical study of the nature of knowledge will compare claims to knowledge in the sciences with those in religion and in everyday life. It will examine the nature of the evidence thought to be relevant in each sphere and in general attempt to draw a clear picture of the concept of knowledge which will enable us to distinguish it from related concepts such as that of belief.

It is sometimes said by philosophers that the aim of a philosophical investigation is to provide a logical map to enable us to pick our way through the confused areas of human thought and action. On this account the philosophical study of social ethics will be an analytical investigation of the main concepts and institutions of social morality with the aim of making clear their meanings, interrelations, and implications. Returning from the metaphor of logical mapwork to that of picturing we might say that social ethics gives us a synoptic picture of the principles and concepts of human thought and action with those of social morality and its institutions in the foreground and the others in the background but still (let us hope) in perspective. In order to develop this account of the nature of social ethics let us compare social ethics first of all with social reform or preaching and secondly with social science.

2. SOCIAL ETHICS AND SOCIAL REFORM

The aim of social ethics, we said, is to clarify the nature of social life in its moral aspect. The aim of social reform is rather to improve the conduct of social life in its moral aspect. Thus the aims of the two activities are essentially quite distinct. To leave the matter there, however, may be misleading, for historically there has been considerable overlap between the two activities. For example, the writings of Marx and Engels have as their basic aim the reform of society, but the method used is that of philosophical analysis. Thus they use philosophical analysis to try to demonstrate the class basis of all morality, and they introduce new concepts, such as 'alienation', to be conceptual tools in their diagnosis of the ills of social life. Hence, although the social philosopher and the social reformer have quite different aims, there may be considerable overlap in the methods used and the content of what is said.

In a similar way, the aim of preaching is to tell people what they ought to do or believe – that they ought, for example, to forgive their debtors. Now a philosopher will not issue moral advice or imperatives in this direct way. But although the aims of the preacher and the philosopher are quite distinct there may be a considerable overlap in the methods of argument used and the content of what is said. For a good preacher will not simply say, 'Forgive your debtors', but will try to explain what is meant by forgiveness; and in so doing he may engage in the conceptual analysis which distinguishes philosophical activity. Sometimes, of course, he may explain what it is to forgive by means of a story or parable, but there is no reason why he should not do so by means of philosophical analysis; and many sophisticated preachers have in fact done so. For example, Bishop Butler actually delivered his *Sermons* at the Rolls Chapel in 1729 with the express aim of altering the practical beliefs and conduct of his congregation. Yet his *Sermons* are a standard text for students of moral philosophy. The reason for this is that he tried to explain to his congregation such matters as the fallacious nature of the view which was common among the sophisticated men-about-town of his time (and our time too) that all actions are self-interested, or the various meanings of the slogan 'Follow nature!' The clarification of these and similar ideas was a necessary condition of recommending the appropriate conduct. Hence, Butler's concerns as a preacher involved him in philosophizing. We may say, then, that although the aim of preaching and of social reform is to alter a man's beliefs and conduct there is no reason why such an aim should not be compatible with, or even in some contexts have as a necessary condition of its adequate fulfilment, the conceptual clarification in terms of which the aim of philosophy is here being characterized.

So far I have tried to show that despite the essentially distinct aims of social reform or preaching and social ethics there may in practice be a blurring of the distinction between the two activities because the one may use the methods of the other. There is, however, a more important reason why the two activities have, historically speaking, affected each other. It is that a clarification (whatever its aim) of ordinary beliefs about social morality will in fact have an influence on social conduct and the principles which govern it. The explanation of this close connection between the theoretical study of social ethics and social practice is that the accounts of social morality developed by philosophers are created from materials which already exist in some form. Indeed, the great moral or political philosopher is distinguished not so much by his ability to create an entirely new set of ideas as by his ability to crystallize those which are already current but lack definite formulation, or which are so general that, like the air we breathe, they pass unnoticed. The result is that when the philosopher produces his system of social ethics we may become explicitly aware of ideas which we have already implicitly accepted. The outcome is a clarification of our moral and political concepts which may well affect our conduct. It is true that the clarification may be described as 'intellectual' or 'theoretical', but a clarification of one's concepts will affect one's views on the facts, and the way in which one sees the facts cannot but affect one's practical conduct in respect of them. Hence, although we can make a sharp distinction between the *aims* of social ethics and those of social reform or preaching, in practice the two types of activity will be more closely involved since the one may involve the methods of the other and have its practical ideas crystallized by the other.

A good example of the close connections in practice between social reform and social ethics is provided by the history of the Utilitarian movement in the eighteenth and nineteenth centuries. Consider, for instance, the utilitarian views of David Hume. According to Hume, the nature of our morality is determined by the constitution of our human nature, and so the key to an understanding of morality is provided by an understanding of the principles according to which human nature operates. But Hume also believed that the most effective method of investigation was that exemplified in the new experimental science of his day, and he therefore attempted to examine human nature scientifically, inquiring why men called certain actions virtuous and others vicious. The answer he gave was that men call those actions virtuous or vicious which afford them a characteristic kind of pleasure or pain, and the kind of actions which provide the characteristic pleasure are precisely the ones we regard as useful to others or to the person himself, or which are agreeable to others or to the person himself. This is Hume's version of the utilitarian principle: it purports to be a clarification of the principles in terms of which we all habitually act. As such it has the logical status of a statement of fact rather than that of a recommendation as to what we ought to do.

But this is not the whole truth about Hume's principle of utility: the clarification of moral ideas which it achieved was accomplished by the selection of certain beliefs and attitudes from the consciousness of the eighteenth-century gentleman and the creation from them of a coherent and persuasive system. In other words, Hume's moral and political philosophy reflected his own moral valuations, which were typical of those of the cultivated men of his times. Further, Hume's utilitarianism acted not only as a clarification of the principles in terms of which the cultivated men of his time in fact behaved, but also as a disguised recommendation that they should continue to behave in this manner. The principles implicit in enlightened eighteenth-century conduct were drawn and erected into a recommended norm. The practical consequences of the utilitarian clarification of moral ideas is even more apparent in the nineteenth century, where we find that utilitarian ideas are used as a theoretical basis for the reform of the criminal code by writers such as Jeremy Bentham and the Philosophical Radicals.

The close connection between philosophical theory and social reform in the history of the Utilitarian movement is only one of many historical examples of the way in which practical social ideals may be affected by philosophical analysis. To say this, however, is not in the least to disturb the central contention that in aim social reform and social philosophy are essentially distinct.

3. SOCIAL ETHICS AND SOCIAL SCIENCE

If we turn now to a comparison of the aim of sociology and kindred social sciences with that of social ethics, we find that the former inquiries aim at providing knowledge about how people in fact behave in social contexts, and about how social institutions in fact operate in given societies at given times. Now a philosopher may (and perhaps ought to) make use of the findings of social scientists, but his own aim is not to add to this knowledge of society and social behaviour. Rather (as we have seen), his aim is to clarify the concepts which ordinary people themselves use in describing and evaluating their social behaviour, and to scrutinize the concepts and methods of the social scientists in investigating ordinary social behaviour. For example, a sociologist might inquire into people's attitudes towards authority and conclude that Group A (factory-workers, say) resent authority whereas Group B (ex-soldiers, say) respect it. Again, a psychologist might inquire into what sorts of upbringing in fact tend to produce certain attitudes to authority, the place of father-figures in developing respect for authority, and so on. But a philosopher is interested in the concept of authority as such. He will therefore ask questions such as: Is authority the same as power? Is there basically only one kind of authority? What does it mean to say that conscience has authority? Social science, then, is factual and additive whereas social ethics, as a philosophical study, is conceptual and clarificatory. Of course, social scientists do attempt to clarify their own concepts, by definitions, etc., but their aim is properly restricted to clarification within the confines of their own and closely related disciplines. If they attempt to analyse their concepts in a wider way and relate them to our ordinary uses of the concepts and attitudes to the world, then they have become to that extent philosophers. And indeed social science, and science in general, does frequently have philosophical moments in it.

Philosophy is especially liable to occur in the *theories* of social scientists. It would be quite misleading to suggest that a social scientist is concerned exclusively with seeking facts about society and social behaviour for he is also interested in explanations of these facts, and to this end he will make use of concepts and laws which attempt to create theoretical understanding of social phenomena. Now it is not always easy to distinguish such social theories from social philosophy, and the reason is simply that large-scale social theories, such as those developed by Max Weber or Talcott Parsons, are, in fact, philosophical in nature. If they fall short of being fully-fledged philosophical theories it is because their breadth is limited. A philosophical theory will aim at being *synoptic*; at relating and explaining human behaviour not simply as a sociological or psychological phenomenon but also as the ordinary agent sees it or as the theologian might see it.

A precisely similar situation arises in physics. If a physicist formulates a theory of considerable breadth and explanatory force – such as the General Theory of Relativity – a case can be made out for regarding it as a philosophical theory; it is not for nothing that theoretical physics used to be called 'natural philosophy'. Some philosophers of course might argue that the presence of quantitative formulations in physics or the use of experimental techniques disqualify it for the title of philosophy. But if one insists on being litigious about these matters (and it must be admitted that some philosophers are *very* litigious about what is or is not properly philosophy), a better reason for ruling out theoretical physics as philosophy would be that it is not synoptic – its aim is to explain the events of a limited aspect of nature whereas philosophical theories tend towards complete world-pictures.

It seems, then, that in distinguishing philosophy from science, and in particular social ethics from social science, we can have no grounds for laying down necessary and sufficient conditions which would enable us to say in any case whether a given study was social science or social philosophy. The best we can do is to say as a rough guide that social ethics, like other branches of philosophy, is non-factual, conceptual, clarificatory and synoptic in its theorizing.

4. SOCIAL ETHICS AND MORAL PHILOSOPHY

It might be objected that if the field of discourse of social ethics is the whole area of social morality then social ethics is no different from moral philosophy in general. For moral philosophy investigates the nature of morality in general, and all morality is social morality.

In reply to this point we might deny that all morality is social morality; it is at least arguable that there is a private or self-referring aspect of morality. It is arguable, in other words, that a person has duties to develop in himself his gifts and his characteristic human endowment. If this is so, then not all morality is social morality. But even supposing we do not insist on this we are not obliged to think of social ethics as being simply another name for moral philosophy. In the first place, there are aspects of moral philosophy which are beyond the scope of social ethics. For example, it is a legitimate concern of the moral philosopher to consider whether moral judgements are better regarded as imperatives or as expressions of emotion or as statements of fact; but such questions lie beyond the scope of social ethics. In the second place, there are aspects of social morality which are characteristically the subject-matter of social ethics rather than of moral philosophy. These are the aspects of social morality in which the influence of institutions is strongest.

It does not, of course, follow from the fact that social ethics is particularly concerned with the institutional side to social morality, that it is not concerned with what might be termed 'face-to-face' relationships; for a relationship can still be face-to-face although it is institutionally structured. The relationships which hold between social worker and client or teacher and pupil illustrate that a moral situation can be both face-to-face and institutionally structured. Indeed, it does not even follow that the relationships which fall within the field of discourse of social ethics are necessarily in all senses impersonal.* Certainly they are impersonal in the sense that institutional concepts may be necessary to describe them, but they may in another sense be deeply personal. For example, the relationship between husband and wife is one which is institutionally structured and in that sense impersonal, but clearly it is in the more common sense a deeply personal relationship. It may therefore be less misleading to say that social ethics, while it is by no means exclusively concerned with, at least concentrates on, those aspects of social morality in which people can be said to act in social roles which are given their structure by social institutions. It will provide analyses of the problems which arise as a result of tensions between the actions of people in their social roles – as *personae* – and their actions out of their social roles – as persons *simpliciter*. Alternatively, we might say that social ethics is concerned with analysing what it is to act in various public capacities and the nature of the relationships between acting in a public and in a private capacity.

5. CRITERIA FOR VALIDITY IN SOCIAL ETHICS

Further similarities between social ethics and social science emerge when we consider that the criteria for a valid theory in each are roughly the same, and indeed are the same as the criteria for validity in any theoretical activity: internal consistency, economy in the use of concepts, and comprehensiveness in its explanatory force within a given area of investigation. Let us examine these criteria in turn and consider how they apply to a theory of social ethics.

* For an analysis of the term 'impersonal' see Chapter 6.

It may seem that no respectable thinker would ever fall into inconsistency in the statement of his theory. Many have done so, however, and the reason is that inconsistency in a large-scale philosophical work is not easy to detect. For example, J. S. Mill maintained in his *Utilitarianism* that the only thing good in itself is pleasure or happiness. He then went on to argue that it is possible to distinguish pleasures qualitatively as well as quantitatively, so that a person might say that one activity is better than another quite irrespective of the amount of pleasure it produced. Now some critics have maintained that Mill was here introducing an inconsistency into his theory because (it can be said) if one pleasure is better than another irrespective of its quantity, then Mill must inconsistently be assuming that something other than pleasure is good in itself. Again, Mill seems to argue that social morality can be completely explained in terms of the principle of utility, but it is arguable that he is making use of other principles such as equality or liberty without realizing that this is inconsistent with the claim of utility to be the sole principle of social morality. I do not here want to discuss whether or not Mill is inconsistent in these ways, but the fact that it is a matter of philosophical debate whether or not a given theory is inconsistent illustrates that the criterion of consistency is by no means easy to apply to a philosophical theory.

The criterion of economy or conceptual simplicity, like that of consistency, is to be found in all forms of theorizing. A striking example of its influence is the preference which Galileo expressed for the Copernican theory of the solar system. It was not uncommon for Protestant historians of a previous generation to maintain that the empirical facts supported the Copernican system and that the bad Roman Catholics refused to look through the telescope and see the evidence, preferring the tongs and bones of Aristotelian metaphysics. The truth seems to be that the best astronomers of the day, such as Tycho Brahe, did not accept the Copernican system because they believed that the empirical evidence was against it. Galileo accepted the Copernican system largely because it explained the known planetary movements with greater conceptual simplicity than its rivals. As he understood it, the Copernican system was a more economical mathematical model – assumed fewer hypotheses than any other – and since he believed as an article of his Platonic faith that the universe was a simple mathematical system he was quite prepared to allow his reason to perform a 'rape on his senses'.

In social ethics the criterion of economy leads philosophers to characterize social morality in terms of as few principles as possible – a single principle if it is adequate. Thus we find that some philosophers maintain that the whole range of seemingly independent moral judgements we make can be reduced to the single principle of utility. We shall later cast doubt on the view that an adequate account of social morality can be provided solely in terms of the principle of utility, but the principles and concepts of the preferred theory will nevertheless be kept to a minimum. In other words, the criterion of economy will be operative.

It should be noted that the criterion of economy is often used in a different way, or alternatively that there is a slightly different criterion of economy to be found in philosophical argument. The criterion to be used in this essay may be called simply a criterion of 'streamlining', but there is a more destructive criterion often called 'Ockham's Razor'. This criterion, wrongly attributed to the medieval philosopher, William of Ockham, is to the effect that there should not exist in a theory 'entities', or concepts with unobservable referents such as 'substance', 'God', or an unanalysable 'ought'. Now in so far as Ockham's razor is simply a methodological recommendation that we should stick to the facts it is harmless enough. But when it becomes a destructive principle to cut out from respectable theorizing whatever is unobservable then it becomes the assumption of a metaphysical view which may or may not be valid, but which certainly requires to be defended with a degree of argumentation quite beyond the scope of a book on social ethics. I shall therefore use the criterion of economy simply as a recommendation to streamline the preferred theory as much as possible, and I shall leave the more dangerous weapon of Ockham in its sheath.

The third criterion for the acceptability of a philosophical (or scientific) theory is comprehensiveness. To be adequate a theory must be able to accommodate, and explain by systematically relating, all the phenomena in its field of discourse. What is the field of discourse of social ethics? The simple answer to this question was given at the start – social morality – but we can now develop the answer by considering an objection, or set of objections, to the whole enterprise of social ethics on the grounds that social morality does not exist.

6. SOCIAL ETHICS AND MORAL DIVERSITY

The objection, stated in general terms, is that moral beliefs are endlessly varied and therefore there is no one phenomenon which can be identified as 'social morality'. Hence, there is no real subject-matter for social ethics. This objection, which we shall call the 'fragmentation thesis', may be pressed in a more or a less radical form. We shall consider them in turn.

In its more radical form the fragmentation thesis might be to the effect that there cannot be anything which is properly called social morality, since morality is a matter of individual choice or preference. If this thesis were sustained it would mean that there would be no field of discourse to be investigated in social ethics since there would be nothing which it is proper to call 'social morality'. Is this radical thesis valid?

An argument in support of it might be based on two premises. The first is that a person must decide for himself what he believes to be morally right if his beliefs are to count as moral beliefs at all. Hence, moral views seem to be in some sense irreducibly subjective or peculiar to the specific moral agent who holds them. The second premise is that what is often called 'social morality' is simply convention or legal obligation. As an illustration of the argument we may mention a distinction often drawn in Parliament between what are called 'matters of conscience' and broader issues of social policy. 'Matters of conscience' concern laws relating to divorce, say, or homosexual behaviour, and they are generally left by the Government to the conscience of the individual Member of Parliament. By contrast, Members of Parliament are expected to toe a line on the issues of social policy. The claim, then, is that whereas there may be some measure of agreement in a society about the general principles of social organization, we find by contrast in moral matters a complete fragmentation of view.

In replying to the objection we are not obliged to deny either premise. It may well be plausible to hold that a view is not a moral view unless the agent has in some sense consciously adopted it as his own. But to admit this is not to admit that moral views within a given society are endlessly variable. It does not follow from the premise that a person must consciously endorse his views for them to count as moral views that everyone will endorse different moral views. Hence, the first premise does not really support the fragmentation thesis.

And neither does the second premise. It is true that many moral rules may be called 'conventions' and that many moral rules are to be found in the legal system. But that goes no way towards showing that they are not *also* moral rules. Indeed, as we shall later argue, it is to be expected that the basic moral rules of a society will also be given legal force. In the meantime, however, we shall assume that in its more radical form the fragmentation thesis can be rebutted.

The less radical form of the thesis does not assert that there are as many moralities as there are individual moral beliefs, but only that there are several social moralities. One might say that in this form the thesis is asserting that the 'fragments' are themselves social moralities; and that there are several of these. This would be, as it were, a 'macroscopic' version of the fragmentation thesis, as distinct from the 'microscopic' version already considered. If valid it would seem to suggest that there is no unified field to be investigated in social ethics.

Now it must be admitted that social morality is not a monolithic organization of global dimensions. Clearly, it exhibits diversity of various kinds. I shall therefore concede straight away that the analysis in this book will be concerned mainly with the social morality of Western civilization and will not pretend to cover the various social moralities of more primitive civilizations. Yet even within Western civilization, it may be said, there are different ideologies, including liberal-democracy, communism, and various forms of the Judaeo-Christian religious tradition. Hence, even in this macroscopic form the fragmentation thesis must still be taken seriously. But not so seriously as to discourage us from proceeding, for three points can be made to qualify its force.*

The first is that despite the undoubted differences which are to be found in the moral principles of persons or communities influenced by the various ideologies of Western civilization there still remains an underlying similarity in the basic principles operative. This is evinced in the fact that their representatives (and those of other ideological groups as well) were able to sign the United Nations Declaration of Human Rights. Certainly, there were important differences in the interpretation of the Declaration, but the fact that they were able to sign it indicates that in very general terms there was a consensus of view on the broad principles of social morality. In other words, I am claiming that there is sufficient structural similarity in the social moralities of the various communities in Western civilization for my present purposes; for these are concerned with the elucidation of concepts and broad principles of social morality rather than with local, racial, or class variation of detail in codes of individual morality.

The second point which can be made in mitigation of the fragmentation thesis in its less radical form will be developed in more detail in the next chapter. At the moment, however, I shall simply state it in general terms. If social morality, whatever else it is, is at least a system of social organization which enables human beings to survive in communal living,

* See also Chapter 5, section 3.

then all continuing social moralities *must* have certain structural features in common. The analysis of this 'must' cannot be completed in a paragraph so I shall postpone it until Chapter 2.*

Thirdly, let us suppose that no single real-life system of social morality exhibits in their purity all the features which I shall attribute to 'social morality'. Let us say for the sake of argument that contemporary social morality in the West is composed of an inconsistent mixture of liberal principles such as Kant's 'respect for persons' principle, of traditional Christian teaching, of the residue of courtly traditions of chivalry, and of folk traditions of morality more ancient and primitive than any of these. It would follow that any attempt to build up a picture of social morality in terms confined, say, to the liberal tradition must be subject to the important qualification that no actual moral or political system would conform exactly to such a description. But it would not follow that a picture of social morality created in terms of predominantly liberal ideas would have no theoretical value, for it is by means of such an idealized picture that we may hope to understand something of the rambling structures which are our actual moral and political systems. To understand the complexity of real-life systems we often need the aid of a simple model; this is as true in social ethics as in science. Hence, if the maximum is conceded to the objection that social moralities vary from one community to another, or even within the same community, it would not follow that the enterprise of clarifying social morality was misguided.

7. SOCIAL ETHICS AND CHRISTIAN ETHICS

It might be objected that social morality so depicted – in terms of the primacy of the individual – is essentially a religious idea. In other words, it might be said that Kant's formula of respect for persons as ends in themselves or the liberal stress on the importance of the individual, are simply

* pp. 26–7.

secular versions of the religious principle of 'Love thy neighbour', and that social morality so understood depends as a consequence on religion. It may thus be argued that any philosophical analysis of morality which does not centre on this crucial feature fails by the criterion of comprehensiveness; it neglects the basis of its subject-matter.

This argument is an important one and merits examination in detail. Of course, the considerations advanced to support this religious interpretation of morality are not all of one kind, and indeed provide material for a book on their own;* but let us investigate four different theses which have sometimes been put forward to sustain the view that morality is in an essential way dependent on religion.

The first thesis we may call the 'historical origins' thesis and it is often put as follows: the idea of the supreme worth of the individual person ('respect for persons as ends in themselves') was originally expressed within Christianity in the form, say, of the precept 'Love thy neighbour'. Hence since the practical force of the basic principle of secular morality had its historical origins in Christianity there is a sense in which the principle can be said to be dependent on religion.

Now there is a certain ambiguity in saying that it is in Christianity that we can find the historical origins of the idea of the importance of the individual, for a search for historical origins can be one or other of two rather different pursuits. In the first place, it can be a search for 'who-got-there-first'. For example, there is often controversy about who was really the first to discover America, or invent the telescope, or introduce the calculus. Interpreted in this way the historical thesis is maintaining that Christians were the first to formulate explicitly the precept 'Love thy neighbour'. In this interpretation, however, the thesis is by no means obviously correct. The Old Testament seems to contain the precept 'Love thy neighbour', and perhaps it existed outside the Judaistic tradition altogether. Certainly, it cannot be

* For extensive discussions of this topic see W. G. Maclagan, *The Theological Frontier of Ethics* (London: Allen & Unwin, 1961), and Keith Ward, *Ethics and Christianity* (London: Allen & Unwin, 1970).

maintained with any plausibility that it is to Christianity that we owe the first explicit formulation of this basic principle. But a Christian can concede this point without much damage; there is indeed something a little comic about fighting hard to defend it.

There is a second interpretation of the historical thesis which is a more plausible and significant one for the Christian position. Here, it is not maintained that Christians were the first to *formulate* the principle, but that they were the first to *stress* it. In Christianity, it may be argued, we first see the principle of 'Respect the individual' or 'Love thy neighbour' exhibited with a proper understanding of what it means. If the first interpretation of the historical thesis can be called the 'who-got-there-first' interpretation, the second can be called the 'who-made-a-thing-of-it' interpretation. And there does seem to be a valid point in the latter. If we concentrate on some at least of the New Testament parables – and forget about the barbarous side of the history of the Christian Church – we can say that Christianity did stress that people ought to be loved and respected just as people.

Now to assert that Christians were the first to emphasize this principle does not in itself establish an *essential* dependence of morality on religion. To establish such a dependence some other premise is required, such as that an insight is always tied to the historical context in which it first appeared. But this additional premise seems very unplausible, as can be brought out in other instances. For example, the Pythagoreans were a sect of mystical mathematicians who discovered many properties of numbers and of geometrical figures, sometimes as a result of their mystical beliefs. But few people would want to say that it is necessary to accept the mystical beliefs of the Pythagoreans in order to accept their discoveries in mathematics. I can believe that the square on the hypotenuse of a right-angled triangle equals the sum of the squares on the other two sides without feeling committed to a belief in the transmigration of souls. It does seem to be true that the gaining of certain insights is sometimes the result of the holding of certain beliefs – there are many illustrations of this point in the history of science – but insights so gained may well be able to stand on their own feet and have no essential connection with their historical origins. The claim, then, is that although much of what is good (and bad) in our present moral outlook had its origins in Christian thinking, such insights do not depend for their validity on Christian beliefs: the insights could perhaps have been gained in other ways, and they can in any case stand up for themselves. The historical origins thesis does not therefore succeed in establishing, on either interpretation, the essential dependence on religion of the principle of respect for persons.

A second thesis may be advanced, however, with such an intention, and this we may call the 'moral practice' thesis: that it is only within the framework of the Christian religion that there is the possibility of putting into practice in any genuine way the conduct required by 'respect for persons'. Once again, we must make this thesis more specific if we are to discuss it with any profit. A crude interpretation – although one with a long and distinguished history, and perhaps even with some justification in the language of the Gospels themselves – is that religion offers a framework of sanctions – rewards and punishments – without which human beings lack sufficient incentive to obey the demands of morally good living. The reply to this claim is the obvious one that action performed out of fear of punishment or hope of reward is not morally good but merely self-interested. We need not discuss this interpretation any further here, although of course it is only fair to note that a Christian may say that what is meant by rewards and punishments in a Biblical context is something more sophisticated than the surface meaning of the language suggests.

A more plausible version of the same point stresses discipline rather than rewards and punishments. It can be argued that success in the moral life requires discipline to make right choice a matter of habit, but that it is only in a Church that one can find the organized, prolonged, and authoritative training that is here appropriate. Now it is probably true that the Church did once provide the main source of moral training, but it was never the only source: the family, the school, and the criticism of one's fellows have always played an important part in moral education. Nowadays, moreover, moral training is increasingly being provided by secular organizations – youth clubs, social workers, women's magazines, and so on. Hence, without in any way belittling the achievements of the Church now or in the past in this matter, we are not obliged by the evidence to admit that the possibility of a genuine moral practice depends in any essential way on the discipline imposed by an established Church.

A different kind of interpretation of the moral practice thesis, and one which brings in religion in a more fundamental way, is based on the alleged need for Divine Grace. It can be argued that by means of prayer the Christian can ask for Divine Grace and that Grace will enable him to lead a life of purer moral quality than that of the unbeliever who is denied, or denies himself, such an inspiration. We may discuss this issue first of all as a matter of empirical fact. Is it in fact true that the Christian leads a life of a significantly purer moral quality than his fellow unbeliever? This does not seem obviously the case. Christianity has produced its saints and heroes, but also its villains and its average church-goers. And the latter, who are perhaps the most relevant here, do not seem to reveal any striking merits which the non-church-goer lacks. What is gained on the swings of chastity is often lost on the roundabouts of complacency.

It may be replied, however, that while the influence of Divine Grace is not to be detected as a matter of empirical fact in moral practice, it is nevertheless operative in the soul. Indeed, it may be said that the whole conception of a better or a worse in moral practice is presumptuous since our actions are all alike sinful, and our only hope lies in asking for Divine Grace to escape from sin. The trouble with such *a priori* arguments, however, is that they are difficult to establish one way or the other. If everything we do is sinful, then the usual distinctions between right and wrong cannot be used to explain the nature of sin; and if the only way of escaping from such radical sin is by the operation of a Divine

Grace, not detectible by normal means, then we have matters that are difficult to argue about. And fortunately we need not, in the context of social ethics, argue them further, except to assert that they do not seem to establish a case for the *essential* dependence of moral practice on religious belief. The most that can be said here for religious belief is that it does seem to help those who have it to withstand hardship and temptation which might undermine their adherence to moral standards. This help is by no means negligible, but it is not such as to justify the assertion of a general connection between religious belief and successful moral practice.

The third thesis which may be advanced to provide an essential link between morality and religion can be called the metaphysical thesis: that morality outside a religious framework is somehow a pointless endeavour. For example, it may be argued that the principle of respect for persons is ultimately meaningless unless it is, say, an expression of the will of God. Again, the point has sometimes been put slickly by saying that the Brotherhood of Man is significant as an ideal only in so far as it implies the Fatherhood of God. This thesis brings out most sharply the difference between the secular and the religious views of life. On the religious view a moral code is the expression of something which goes beyond and transcends human affairs and human beings, while on the secular view a moral code may be said to have a function such as that of organizing people in society on a harmonious and co-operative basis. A religious view of morality need not deny that morality does have this function as well, but on this view such a function is incidental, or at least subordinate, to the religious significance of morality. If we are to give a functional interpretation to morality, the religious world-view will see the function as one of disciplining and in general of preparing the soul for its final communion with its Maker.

In the end it is not possible to provide conclusive arguments for or against such a position. What are involved are fundamental attitudes towards human life and its place in the whole scheme of things. In the end there may be several possible world-views which may be taken up, but there is no neutral way of deciding among them. One of them may be true, of course, and the others false, but we are in no position to decide which one is the true one. We can, of course, evaluate a metaphysical system in terms of its internal consistency, its comprehensiveness, or its ability to account for all the basic factors of our experience, and its economy in the assumptions and concepts it requires to do this. Such factors can lead to the elimination of many metaphysical systems which have been built up by philosophers in the past, but we shall be left with several systems, including the religious one, and no way of deciding truth or falsity. Hence, to set out the religious context of morality will convince a religious believer of the essential dependence of morality on religion, but it will not convince a person whose metaphysical views are different. What may be done, however, is to show that the non-religious view of morality, the view we have characterized by the principle of respect for persons, goes deeper than a religious person may at first believe. To try to bring this out let us examine an example which cannot be regarded as unfair from the religious point of view – the parable of the Good Samaritan.*

The point of the parable is to answer the question 'Who is my neighbour?', and the answer seems to be that my neighbour is not just a person with whom I happen to be in a special relationship but is any other human being needing help. Now if one ought to love one's fellows simply as human beings and not just because they happen to share one's race or religion or politics, then it follows that to say, 'He is a person in distress', is to provide a *sufficient* reason for action. But if it is once conceded that a sufficient reason for action is provided by the fact that a fellow creature is in distress, then it has also been conceded that morality is not essentially dependent on religion. And this is the main point to be made in a context of social ethics. It may be argued that an *additional* reason for helping one's neighbour is provided by the fact that one's neighbour, one's brother, is also a son of

* *St Luke's Gospel*, Chapter 10, 25–37.

God, or that God commands us to love our neighbour. Now these may well be additional reasons, as we may have more than one reason for doing the same thing, but my claim is that these reasons are only contingent on the main reason – that one's neighbour is in distress. And if this reason is sufficient we have an autonomous social morality which does not essentially depend on any sort of religious belief. It is arguable that this is the message of the parable of the Good Samaritan.

Advocates of the fourth thesis might concede all the arguments so far propounded and agree that social morality is autonomous in the sense that its concepts are not reducible to those of religious belief or any other sphere of human interest. They might go on to point out, however, that there remains a way of looking at social morality which permits us (although it does not compel us) to see it is as a Divine phenomenon. This possibility exists provided we do not think of the religious significance of morality as being distinct from its moral significance. In other words, the claim is that a moral commitment can itself be regarded as a religious commitment; in accepting the one commitment a person is *ipso facto* accepting the other as well. We can call this the 'identity thesis'.

It is easy to confuse the identity thesis with the metaphysical one previously discussed. The metaphysical thesis portrayed the religious significance of morality as being *additional* to its moral significance; morality was thus said to point beyond itself and to exist for the sake of some further and higher end. The identity thesis, on the other hand, depicts moral significance as being itself identical with religious significance.

As a further elucidation of the nature of this thesis consider the parable of the sheep and the goats, which gives it scriptural authority.* In this parable God is described as a King passing judgement. He says to the righteous, 'I was an hungred and ye gave me meat: I was thirsty, and ye gave me drink: I was a stranger, and ye took me in . . .'. The

* *St Matthew's Gospel*, Chapter 25, 31–46.

righteous are puzzled by this and ask when they did these things. God replies, 'Inasmuch as ye have done it unto one of the least of these my brethren, ye have done it unto me.' The point here is not that the hungry were fed *for the sake of* some other religious value; they were fed for the sufficient reason that they were people who needed feeding. The point of the parable is rather to assert that this total moral commitment to the needs of other people is also, in virtue of that very fact, a religious commitment. The identity thesis, then, does not so much reduce morality to religion as reduce religion to morality (and other serious commitments).

An adequate discussion of this thesis is beyond our scope. Its philosophical roots are in a view developed by Kant, who interprets the experience of being under categorical moral obligations as an assurance of the Divine ordering of things, and it appears in a slightly different form in the work of some modern theologians such as Paul Tillich or John Robinson. There are certainly many theological and philosophical difficulties in it, but it may be the most satisfactory way in which a person with religious beliefs can approach social morality; for it enables him to take with utter seriousness commitments to the needs of other persons as such, and yet it does not prevent him from tying in such commitments with his religious views. On the other hand, a person who has no religious views does not emerge from this analysis as one who lacks anything of significance for a commitment to social morality.

The four theses we have discussed by no means exhaust the possibilities of religious interpretations of morality, but the discussion may have served to indicate the lines along which the autonomy of social morality, interpreted as an expression of the principle of respect for persons, may be defended while still leaving room for a meaningful assertion of religious views by those who have them. The arguments which have been formulated in the course of the discussion are not to be seen as an attack on religion, but simply as a resistance to the engulfment of the language of social morality in that of religious discourse.

8. CONCLUSION

This chapter has been introductory in respect of its content but not in respect of the types of argument used. It has been introductory in that I have attempted to explain the nature and aims of social ethics. Thus I have tried to show that they must be distinguished from those of social reform and social science; that the criteria for validity used in social ethics are the same as those used in other theoretical disciplines; that despite arguments which make use of the idea of moral diversity there is still something which can significantly be called 'social morality'; and that it can be regarded as an autonomous phenomenon despite its many connections with religion. But the arguments of the chapter have been fully-fledged philosophy from the outset. Indeed, it is not possible (logically) to say in any detail what philosophy is without engaging in it, and it is not possible (psychologically) to understand the activity without practising it. Let us now direct the activity to the positive task of clarifying the structure of social morality and its institutions.

2

The Principles of Social Morality

I. THE CONTEXT OF MORALITY

Philosophers in the seventeenth and eighteenth centuries often began their accounts of morality or politics with a description of man in a 'state of nature'. These descriptions were sometimes conceived by their critics, and perhaps by their authors as well, as being historical or anthropological accounts of man as he was before society, or at any rate political society, had been formed. As such they are easy to criticize for the lack of empirical evidence on which their generalizations are based, and in general for their naïvety. The result has been that such prolegomena to moral and political philosophy have been omitted entirely from recent serious writings. Yet such accounts of man in a state of nature did fulfil a function in philosophy which has not been taken over by any more sophisticated substitute. The function is that of making clear what is the real point in having a system of morality or of politics. The accounts brought out in a picturesque way the strengths and weaknesses of human beings and the nature of the human environment, and the reader was the better able to appreciate the deficiencies in the human situation which moral and political systems to some extent remedy, and to understand how from the raw material of human capacities and the natural environment a system of moral or political regulations can be constructed. If he has stated fairly and fully the nature of human beings and their situation a philosopher is less likely to distort anything in his account of morality or politics. Let us therefore try to describe in very general terms the nature of human beings and the predicament for which morality is the partial solution.*

The description of human nature we shall give will consist in a series of empirical truths about persons and of inferences drawn from these truths. No substantial evidence for the generalizations will be offered because the empirical truths are not intended to rise much above the level of truisms, and the evidence for truisms, where it can be provided at all, will tend to be less well founded than what it is designed to support. Thus, the account to be given will not amount to anything that can be called a psychological or an anthropological theory about human nature or society. What, then, is the point of propounding a set of admitted trivialities about human beings?

Firstly, it will prevent a theory of morality from being based on a one-sided view of a person. The implausibility of many philosophical accounts of morality – such as the more extreme forms of Egoism or Emotivism – would have been obvious had their authors begun with a list of truisms about human nature. It will, secondly, bring out the close link between the kind of nature we have and the kind of morality we have. To say this is not to say that morality can be deduced from the facts of human nature, but rather that we accept the kind of morality we do because we are the kind of people we are; that any plausible account of morality must have close links with an account of what people and their environment are like.

It must always be remembered, of course, that some philosophers have asserted that we ought to act against the tendencies of human nature or even that we ought to bring about its destruction, and that such assertions, even if we disagree with them, make perfectly good logical sense. Hence, any attempted definition of 'ought' which logically prevents the making of these assertions must be rejected.

* The account I shall give is based on H. L. A. Hart, *The Concept of Law*, Chapter ix (London; O.U.P., 1961). Hart is exceptional among modern philosophers in that he makes use of a philosophical anthropology.

The point is simply that *because* people and their environment have certain obvious characteristics they will *tend* to accept certain forms of social organization. The 'because' here is to be analysed causally rather than logically; it is not that 'ought' *means* 'what pertains to social survival', but only that most people in fact desire this. That was the main point behind the colourful accounts of man in a 'state of nature' to be found in the writings of earlier philosophers, and we shall now try to state (in a more sober way) some of the obvious truths about people and their environment contained in these accounts.

The first obvious truth about human beings is their *lack of self-sufficiency*. They are vulnerable in an environment which is in many ways hostile, and thus require protection of various kinds. But a man cannot always protect himself, nor can he always satisfy his own needs for clothing or food. We may say, then, that human beings have basic needs of an obvious sort which prevent them from being individually self-sufficient since they cannot satisfy many of these needs without the aid of other men.

By and large it is true that these needs are felt by human beings as wants. Biologically speaking we may have a need for food, but this is experienced in the human consciousness as a desire for food. Of course, some people may not always feel their needs as wants – pathological conditions are possible, and people can be distracted from their needs by other considerations – but in general it is through the conscious experience of needs as wants that survival is possible for human beings. Our wants extend beyond our needs, however, and express human interests in vast numbers of objectives other than bare survival. Many of these wants are non-material. For example, human beings seek personal relationships, pursue knowledge, enjoy art, and express themselves in religion. How far human wants can be said to extend, whether there is any empirical limit to the objects a person can be said to want, is a philosophical question which it would be inappropriate to raise in a context of truisms. But, as far as our ordinary views are concerned, it is surely true to say that the objects of human wants are exceedingly varied.

A second trivial truth about human beings is that they have (in Hume's phrase) a limited benevolence or a confined generosity. It is obviously the case that people are often out for themselves (and why not?) but it does not follow that they are *always* out for themselves.* One person will sometimes help another in misfortune; and in humdrum ways the average person does this every day. Of course, if the assistance involves some great sacrifice then not everyone will oblige without some assurance of reward in this life or the next. But we need not deny this. The only point it is necessary to assert is simply that people can, and often do, act in benevolent ways where this does not involve them in great hardship. Biologically speaking this is to be expected, since human beings are gregarious animals and their biological inheritance, developed in family life, will make for partial benevolence in their social relationships.

A third obvious feature of human nature is that people are approximately equal in power. Of course, there are large variations in intellectual ability, physical strength, ability to win friends and influence people, and so on. But despite all this it seems to be correct to say that no one person is likely to be so much superior to everyone else that he could hope to subdue others for any length of time. Even the strongest must sleep, and then he can be hit on the head.

A fourth obvious feature of human beings is that they are limited in understanding, knowledge, skills, and so on. To say this is not in the least to belittle the great achievements of the human intellect, but to make the point that these achievements very often depend on human co-operation. As Newton said, 'If I have seen farther [than other men], it is because I have stood on the shoulders of giants.' The development of human knowledge, then, requires the division of labour, and is cumulative. Human beings, as a species, may be developing in knowledge and understanding all the time, but it remains true that the knowledge of any one person is limited. Moreover, the limitation in knowledge is not simply

* The theory of egoism will be investigated in more detail in section 3 of this chapter.

a theoretical matter; the knowledge of conduct which is presupposed in everyday life also depends on the accumulated wisdom of mankind (to use Mill's phrase). The consequences of actions are not always obvious, but an individual person has access to the experience of the human race in rules of conduct.

The environment, fifthly, is an important factor in shaping the form of social organization which people tend to accept. As a result of their basic needs, human beings, as we have seen, require food, shelter, clothing, etc. Now clearly these do not exist naturally in limitless abundance, even in the friendliest of environments, but have rather to be grown or built or manufactured in some way by human labour. In other words, there is a scarcity in the commodities which supply human needs and wants. In view of these facts, and others, human beings require for their survival some form of the institution of property; and if there is to be such a thing as the institution of property there must logically be sets of rules creating it and defining the conditions for its transfer, protection, etc. Thus the rules of property must tell us first of all what constitutes ownership, and how some forms of property may be sold or exchanged for others. And, if one simply considers such concepts as 'trespass', 'buying', 'selling', 'exchanging', 'renting', 'hiring', and a host of others, one thing at least is immediately clear: that there is a need in any form of human organization with property (and perhaps all forms of human society do have some form of the institution of property) for a highly complex set of rules.

Social and political theorists have for long been aware of the importance of these features of human beings and their situation for determining the nature of social organization. Awareness of them is to be found in Plato's *Republic*, for example, in Natural Law philosophers and in Empiricists such as Hume. Some theorists, having noted the features, or some of them, go on to suggest unrealistic or unnecessary doctrines of social contracts between people to form societies, or between societies and rulers. It seems more convincing to say, as Hume does say, that experience of family life leads men to understand the usefulness of certain forms of organization, and that these forms of organization become ever more complex as the social unit grows in size and diversifies. Let us now consider one historically important theory which attempts to explain the complexity of the rules which have grown up to mitigate the effects of these features of human beings and their situation. The theory we shall examine is that of utilitarianism, and we shall consider how far it can account for the diverse and complex nature of the rules which have grown up in response to human needs.

2. UTILITARIANISM

If it is accepted that human nature and its situation are roughly as they have been described we can see that it is plausible to say that human beings and their actions should be governed by some very general principle which mitigates the disadvantages of their situation. And this is precisely what the utilitarians are doing when they assert that actions are right or wrong in so far as they tend to produce the best possible consequences for the majority. Some utilitarians, particularly those of the nineteenth century, tended to interpret 'best possible consequences' in terms of happiness, and the analysis of happiness gives rise to problems. Fortunately these problems need not be raised, provided we retain the more general formulation of the principle of utility which does not mention happiness at all.

As a first attempt to express the basic principle of social organization there are clearly many merits in the utilitarian formula. In the first place, it is a formula which we all do in fact employ when trying to persuade someone to act or refrain from acting in a certain way. 'Think of what will happen if you do X' or 'The results will be highly beneficial' or 'The consequences were disastrous' are all familiar types of utterance in practical contexts.

In the second place, the formula does not say that the consequences must be *good*. Sometimes, on the one hand, an action may produce no good consequences at all, or at least very few, and still be right. For example, a patriot may have the choice of inciting people to rebellion or allowing tyranny to continue, and while one decision may be right perhaps neither will produce good consequences. An action can be right but be the lesser of two evils. Sometimes, on the other hand, an action may produce good consequences but still fail to be right. Most actions are productive of some good consequences, one way or another. Even the decision to engage in war may produce good consequences in the form of the development of brain surgery and other forms of technological or scientific research which have beneficial consequences for human beings. Thus, the production of good consequences is neither a necessary nor a sufficient condition of the rightness of an action; and this is reflected in the utilitarian formula that actions are right if they produce the best possible consequences.

But the formula also stresses that the consequences must be the best possible for the majority, and this is its third merit. It is saying that actions are right not simply because they are in the agent's own interests but because they are in the interests of the majority. Now the claim that it is a merit of the theory to assert this needs some defence, because there are theories to the effect that actions are right or wrong in so far as they are in the agent's own interests. Such theories, generically called 'egoistic', are so important historically, and make such an appeal to the cynical streak in all of us, that it may be worthwhile to disentangle some of the strands of argument involved in them.

3. EGOISM

The theory of egoism is a rival to utilitarianism as an account of right action in that it asserts that actions are right or wrong, or are obligatory, in so far as they are in the agent's own long-term interests. This view of what people morally or rationally *ought* to do must be distinguished from two other strands in the theory of egoism: that the agent *does in fact* always pursue his own interests or pleasure, and that he *can do no other* than pursue his own interests or pleasure. Before examining rational or ethical egoism let us look at these latter views.

In the history of thinking about human nature it is possible to find two opposed accounts, one optimistic and one pessimistic. According to the optimistic view human beings are basically benevolent or generous by nature. It is true (the exponents will admit) that people are often in fact selfish in their actions and policies, but this is to be explained by bad social conditions or institutions which bring economic pressures of a dehumanizing sort to bear on people, or which corrupt them with bad education or otherwise alienate them from each other. If people were once freed of the artificial conditions and false values of our present social organization (it is claimed) then their inborn altruism and love of each other would again come to the surface. This optimistic view of human nature – to be found in a variety of forms in thinkers as diverse as Francis Hutcheson, Rousseau, and Marx – sees society as corrupting and exploiting a basically good and benevolent human nature. Human nature does not need improvement; but the social conditions which corrupt it certainly do.

According to the other view of human nature (the pessimistic one) human beings are radically selfish by nature, out for their own ends on all occasions. It is true (the exponents will admit) that people often seem to be acting in an unselfish or even in a benevolent way, but this is only a façade; behind the outward show lies the incurable egoist. The point of social organization on this view is to achieve a compromise between the egoistic policies of each individual. Some thinkers, Thomas Hobbes, for example, take such a gloomy view of human nature that they think man's only hope of salvation lies in a political society governed by an absolute and arbitrary sovereign. A somewhat similar line is taken by some theologians, who see man as so ensnared by sin that his only hope of salvation lies in Divine Grace, which he may ask for but cannot obtain by his own efforts.

It would seem that each of the opposed views of human nature is an exaggeration. The view that human nature is basically good may have appeared plausible at certain stages in history, and thinkers who wanted to blame, alter, or reform political and social institutions would have a tendency to adopt somewhat idealistic views of human nature. But nowadays even the most zealous of social reformers would hesitate before attributing all human ills to the institutional environment. It does seem to be true that people have a *tendency* to pursue their own interests at the expense of those of others, and that no amount of education or social reform will ever eradicate this. But, hopefully, the pessimistic hypothesis is also a distortion. Certainly, as Kant puts it, the 'dear self' is always turning up, even in the most altruistic-seeming actions, but it does not seem to be plausible to say that people *always* pursue their own interests; occasionally one person can act against his own best interests or can want the good of another as his own. This is the 'trivial truth' expressed in Hume's view that human beings have a 'confined generosity' or 'limited benevolence'.

It should be noted that, even supposing it is true that people always in fact pursue their own interests, this cannot support the theory of rational egoism – that people *ought* to pursue their own interests. The 'cannot' here is a logical one, for from a premise containing only a statement of what *is* the case – that people in fact pursue their own interests – it is logically impossible to deduce a conclusion about what ought to be the case – that they ought to pursue their own interests. Hence, the pessimistic view of human nature seems to be false on its own account and to have no logical bearing on the theory of ethical or rational egoism.

The third strand to be found in egoistic views – that people can do no other than pursue their own interests or pleasure – is of quite a different logical status from the second but is often introduced quite imperceptibly into a discussion of it. For example, one person will assert that some apparently heroic or self-sacrificing action was in fact simply furthering self-interest. Another will deny that this was so, perhaps supporting his denial by providing documentary evidence to the contrary. Now if the first person persists in his cynical assertion despite the contrary documentary evidence it may be that there has been a shift in the logical nature of the assertion. What began as the claim that the agent *was in fact* pursuing his own interests has become a logical claim, for the cynic will not accept contrary evidence as affecting it. The evidence that a person is acting for his own interests (or pleasure) has become simply the fact that he is acting. The end of all action has become *necessarily* an agent's own interests. But it is one thing to observe human action and be cynical about its motives, and quite another to make one's cynicism true by definition. The third strand seems, then, to be a matter of logical stipulation (from which nothing of substance can logically follow) masquerading as tough-minded psychology.

We may add here that even if the third strand were accepted as a psychological thesis it would by no means support rational or ethical egoism. Indeed, it would become obscure how a system of social organization could exist at all. For if there is only one type of action an agent *can* perform it is not at all clear in what sense he can be told that he *ought* to perform it; and the point of social organization is to tell people what in general they ought to do.

It seems, then, that the second and third strands in the theory of egoism are in themselves of doubtful validity and uncertain logical status. But even if they were acceptable, they logically could not support the main strand in egoism, that actions are right or wrong to the extent that they produce the best possible consequences for the agent himself.

But is ethical or rational egoism (without the specious support) a plausible account of the nature of right action? The answer is that it is not, but it *seems* to be so because it does have some plausibility as a different kind of theory, a theory of the *motivation* which might lead a person to accept the rules of a utilitarian or other form of social organization. If we recall the trivial truths in terms of which we characterized the human situation it is clear that an individual, left to his own devices, could hardly survive at all. He has therefore a strong motivation of self-interest for accepting some form of social organization. This will involve constraints on his conduct, but the benefits to him are obvious, provided also that others keep to the rules of the organization. We can therefore say that the individual has an interest in being a member of a social group and in ensuring that it continues to exist. But to say this is not to say that the actions of any given individual are *right* because they further his own individual interests. The ideas of rightness or duty or obligation logically depend, as we shall see, on the idea of rules, and rules are concerned with organizing *group* interests. Hence, whatever plausibility egoism may have as an account of motivation – of what it is rational or prudent for a person to commit himself to – it is not plausible as an account of the basic principle of social organization. It is a merit of utilitarianism that it avoids the pitfalls of egoism and stresses that moral and political rules of social organization are mainly concerned with creating the best possible consequences for the majority in a society.

4. RULES

But what constitutes the 'best possible consequences for the majority'? In general terms, we can say that the best possible consequences will be brought about if there are sets of rules creating social harmony and social co-operation. And we would expect to find rules under these general headings if human beings and their situation have been at all accurately characterized in terms of the features we have already discussed. We would expect that rules would develop forbidding assault or the making of malicious comments on others, for such rules help to prevent disharmony in society. Again, we would expect to find rules which encourage various forms of social co-operation, because such rules would mitigate the effects of human lack of self-sufficiency. These are the rules bringing about the division of labour, exhorting people to help others in the infinite varieties of distress to which human beings are liable and in many other ways making for social utility by directing co-operation in social life.

There is another way of looking at social rules. They can be classified into empirical generalizations about what is likely to happen when people act in certain ways and which express the 'accumulated wisdom of mankind' on the consequences of action; and into what we may call 'institutional rules', or rules defining or establishing a set of social arrangements such as 'property', 'promises', 'lending and borrowing', and so on. Let us consider these two classes of rule in more detail.*

It has already been pointed out that people as individuals do not have the necessary capacities to work out the consequences of all their actions. The experience, understanding, and knowledge of any one individual or group of individuals is limited – this, as we saw, being a basic fact of the human situation – and so they cannot always work out the calculus of consequences for themselves, let alone for society as a whole. Moreover, supposing they could calculate the effects of their actions, they do not always have time to do so before acting; people are often obliged to make up their minds quickly on what is right or wrong, whereas it would take time to go into the probable consequences of actions. Hence, the argument runs, rules have grown up which express the accumulated wisdom of mankind on the consequences of action.

Mill compares this function of moral rules to that of signposts or the Nautical Almanack. We have signposts to guide us across country which may be unknown to us. Again, the sailor does not need to make his calculations at sea but goes to sea with his calculations already made for him in the Nautical Almanack. Similarly, we go across the sea of life with the consequences of our actions already calculated for us in moral rules. Thus, the utilitarian sees moral rules as being rules of thumb or ready reckoners which compensate for the deficiencies in the experience, knowledge, and understanding of the individual person, and the lack of time for calculations which is a feature of ordinary moral situations.

* The account of moral rules given here is a shorter version of R. S. Downie and Elizabeth Telfer, *Respect for Persons*, Chapter II, section 3 (London: Allen & Unwin, 1969).

A utilitarian can also stress a second main function performed by moral rules. In addition to the human deficiencies for which 'lightning calculators' can compensate, human beings are also deficient in altruism. They therefore require the kind of social coercion provided by rules. The existence of rules makes generally known the types of action which are conducive to majority interests. Moral rules as a result come to be backed by sanctions. Where the rules are basic to continued stability they may be incorporated in a legal system and supported by the threat of legal sanctions, and they will in any case be backed by social approval and disapproval. These factors encourage adherence to the rules in two ways. In the first place, many people will follow certain behaviour patterns simply because there is a generally accepted rule urging conformity. Habits of obedience and the desire to conform to established customs are deeply engrained in our society – indeed, in most societies – and are often sufficient in themselves to overcome temptations to depart from the custom. Many people, in other words, will conform to a rule simply because it is a rule and regardless of its content and sanctions. In the second place, where habits of obedience are less strong, or the temptation to deviate is more strong, the sanctions attached to rules may operate and encourage people to conform. There is therefore a second main function of moral rules which admits of a utilitarian interpretation.

Now rules of this first type are, as we have said, empirical generalizations about the results of types of action. Actions are not made right or wrong because of the existence of these empirical generalizations forbidding or enjoining them; rather we make these empirical generalizations because we have learned from experience that certain types of action have consequences which are liable to be good or bad. For example, 'One ought not to talk about the invitations one has received to parties' is a rule based on the fact that such discussion can be misleading, hurtful, or creative of jealousy and ill will. It is not that any one case of such tactless talking is wrong *because of* the rule, for in judging the action as wrong in a given case one need not refer to the rule at all; it is simply that the rule provides safe moral guidance if one is in doubt. Again, it is often tiresome, irritating, or hurtful to make a joke to someone about his appearance, and consequently one may make it a rule to avoid this line in witticism. But such a rule merely expresses what experience teaches; the wrongness of hurting a person's feelings in this way does not depend on the existence of the rule. We can therefore say that there are rules which have the status of 'wise saws and modern instances' and the rightness or wrongness of the actions covered by the rules is established independently of the rules. Moreover, since rules of this type are only generalizations, they readily admit of exceptions – cases where an action of a kind which normally has bad consequences will have good consequences, and vice versa. Where exceptions do occur we may abandon the rule without a qualm, since the rightness or wrongness of the action does not depend on it.

Other rules, however, have a different logical status. They are not empirical generalizations about the consequences of actions; rather they lay down the obligations inherent in some institution which is artificial in the sense that it owes its existence to rules. This is the case, for example, with rules that one should keep one's promises and with rules concerning property, such as 'Do not steal' and 'Pay your debts'; there would be no such thing as promises, stealing, or debts if there were not rules of conduct defining them. The rightness of promise-keeping or debt-paying and the wrongness of stealing can therefore be said to depend essentially on the existence of rules, and thus to be artificial, in that it owes its existence to the institution which defines the practice. The utilitarian argument is that the operation of such institutions may on the whole be in the general interest even although individual instances of promise-keeping and debt-paying (say) may seem to be against it. We undermine the institution if we raise the question of interest in every case; the rules of the institution must be applied to preserve the institution. In this way the rules of institutions differ from empirical generalizations about the consequences of individual actions.

The institutions of property and promise-making, with the rights and correlative duties bound up with them, are institutions in a rather vague and weak sense; the rules are not necessarily written down but are agreed upon by a kind of tacit consent. In any society there will be many other institutions, some of a far more formal character, possessing the same general features: namely that they set up rights and duties which seem to override the claims of the general welfare. The justification for awarding these special rights and duties lies in the social utility of the institutions which give rise to them. Thus it will always be possible to ask whether an institution, or the current form of it, does promote the general welfare. For example, take the institution of the family. We might ask whether some different conception of the family – more closely knit, wider in scope, etc. – might better promote the general welfare, or whether we would do better to abolish the institution altogether, as in Plato's *Republic*.

These facts about institutions explain many judgements we make which at first sight go against the principle of utility. For example, we do not feel that a man is doing wrong in confining the area of his service to others to a small circle such as his immediate family and friends, although it might seem that the general interest would be better served if he did not merely consider a privileged group. The reason why we do not think it wrong is now clear, and it is a twofold reason: a man should consider his family in preference to others because he has special duties to his family and they have special rights against him; a man is justified in 'taking on' these special duties because the institution of the family is (we may suppose) more conducive to the general interest than any viable alternative.

A great many of the ordinary person's duties are institutional duties of this kind; the scope left to him for direct consideration of the general welfare (such as by charitable contribution of money or service) will be small. A great deal more will be said later (Chapter 4) on the analysis of an institution, but in the present context it is sufficient to note the important difference between moral rules empirically and institutionally based.

So far I have outlined the most obvious features of human nature which make a system of social morality both possible and desirable, and have suggested in very general terms how the principle of utility might help to remedy these defects. The theory of utility provides, then, a first approximation to an account of social morality. There are various respects in which this approximation might be criticized. For example, despite the introduction of the concept of an institution, we cannot so far draw a distinction between morality and legality or in any way distinguish between moral and political organization; the convenient term 'social organization' obscures the distinction. This defect, and others, will later be remedied, but in the meantime let us consider three important principles which are not included in the first approximation: equality, liberty, and fraternity.

5. EQUALITY, LIBERTY, AND FRATERNITY

To base a social organization solely on the principle of utility is to have a set of rules concerned only with maximizing interest. The principle of utility, that is to say, does not by itself suggest how the benefits in the society should be *distributed*. Thus, a society might achieve a high degree of utility by having a slave or otherwise underprivileged section of its population to perform the menial tasks. Yet even if there were an overall prosperity we might still criticize such a society on the grounds that it was unjust or inequitable. It therefore seems that we judge a society not only in terms of the amount of benefit but also in terms of the way the benefit is distributed; and we may well prefer a society where benefit is reasonably equally distributed to one where the total benefit is higher but the disparities between the best off and the worst off are great. To say this is to say that in a society which is morally wholesome the principle of utility must be supplemented by the separate principle of equality.

Utility and equality do not exhaust the basic moral principles of a social organization. To bring this out, consider T. H. White's account in *The Once and Future King* of the principle which King Arthur found when, as a boy, he was enabled by Merlin's magic to visit an ant-heap: everything which is not forbidden is compulsory. A society based on such a principle would certainly have a high degree of utility and equality, but we should still say that it was morally defective. The moral defect consists in the absence of the value of liberty; we feel that unnecessary interference with the liberty of people to do as they want is wrong, that we ought to be free as far as possible to pursue our own aims and fulfil our own aspirations. Hence, an adequate account of social organization must make a place for a principle of liberty, as well as for those of utility and equality.

It is clear that the relationships between these three principles will be complex. For example, it is obvious that if there is complete liberty in the pursuit of objects of interest then some people will be able to subdue others and to obtain much more for themselves. The principle of liberty must therefore be combined with that of equality, to produce the conclusion that each person has an equal right to pursue objects of interest. In other words, he may act at liberty to the extent that every other person may do likewise. But the equal pursuit of objects of interest may still produce clashes and there is therefore a need for utilitarian rules of social harmony to minimize such clashes. Moreover, for maximum satisfaction of interests some degree of enforced co-operation is necessary, i.e. a further set of rules based on utility must be enforced. For example, if a citizen is to have certain medical benefits – if he is to be materially or effectively, as distinct from simply formally, free to seek medical benefits – then there must be rules of social co-operation governing, say, taxation. But taxation laws will interfere with the individual's liberty to do as he wants with his own money. Yet we might allow that such interferences with liberty are justified by utility. Again, we might allow that it would be legitimate for a government to take a farmer's land for a reservoir which was essential to supply water to a large urban area. We should certainly insist that he receive compensation, but no amount of financial compensation might make up for the loss of a traditional way of earning a living. In such a case there would be an offence against the principle of equality – for the farmer is being singled out for unequal treatment – and also against that of liberty – for the farmer's liberty to pursue an occupation of his own choosing in a geographical area of his own choosing is being removed. Yet we might still judge that in this case the principle of utility was overriding.

It is, of course, debatable how far individual liberty can be restricted in the name of utility. To put the question in a different form, assuming that restriction of liberty is justified by the utilitarian principle of social co-operation, we may query how far that principle can legitimately be enforced. To raise this question is in fact to raise the whole issue of the moral basis of the welfare state. It is clear that there should be rules enforcing co-operation on matters such as health and hygiene, but how far beyond these matters a government or official body is justified in imposing further restrictions is a question which raises the issue of socialism versus *laissez-faire* individualism. It can be argued, for example, that the formal freedom to enter a university or to be employed is of no value. Freedom is worth having only if it is material, and it becomes material freedom only when it is guaranteed by the State. The price of this guarantee, however, is a degree of restriction with which anyone living in a socialist state is familiar. The price may be worth paying, but the point is arguable – and the fact that it is arguable indicates that liberty has a value in its own right which cannot be reduced to that of utility.

Again, it is not clear in certain difficult cases how to balance the claims of equality and utility. Let us imagine a wartime situation in which the allies can gain a great military advantage which will shorten the war and save millions of lives by means of a mission involving the certain death of those carrying it out. In such situations volunteers may be called for (and indeed are often forthcoming), but let us suppose that there are none, or that for reasons of security it is impossible to divulge the real nature of the plan and its outcome. In such situations it is arguable that the commanding officer is morally justified in selecting certain men for the fatal mission. But this does not mean that utility can completely overrule equality. For the men selected may still retain a right of equality – the right to equal consideration with others when the time for selection comes. In other words, the commanding officer would be morally unjustified in simply assuming that certain men just did not count at all and could therefore be used for this task. Each man, as an end in himself, retains an equal right to consideration with every other. This may be a poor right and a thin right, but it is an absolute one; and for that reason equality must be included with utility in a list of basic moral principles.

We can see that the deficiencies of an account of social organization in terms of the single principle of utility are to some extent remedied by the inclusion of additional principles of equality and liberty. But a defect of another kind may be identified if we say that so far the analysis has been concerned exclusively with *rational organization*, whereas there seems to be a dimension of social morality which involves social feelings and emotions. These are the ties generated by kinship, the consciousness of a common religion, customs, language, traditional ways of earning a livelihood, and traditional loyalties of all kinds, and more generally the sense of being part of a broad cultural tradition.

The concept which may be used to sum up this emotional side to social morality is that of fraternity. There is a certain ambiguity in the concept of fraternity, but in its most important sense a principle of fraternity suggests a motive for action stemming from devotion to some community to which one belongs. In so far as people in a community really are imbued with a feeling of fraternity, they will not see restraints on liberty in the general interest as restraints – in fact they will not be such, but rather ways of bringing about what the individual wants most. This is the spirit which communism hopes to foster in its members. From motives such as these, people in a democracy choose to spend time in the service of others; and whereas it would be an infringement of liberty to make them do so, it is an exercise of liberty if they choose to do so. It is fraternity in this sense which comes nearest to remedying the deficiency in the account of social morality which contains only the principles of utility, equality, and liberty.

The aspect of social morality I wish to identify by means of the concept of fraternity is often made in a slightly different way by means of a distinction between a *Gesellschaft* and a *Gemeinschaft*. The distinction was introduced by the German sociologist Tönnies,* and it can be explained if we translate the distinction as one between an organization or an association, such as a business or other commercial or official enterprise, and a community, such as may be found in a country district which has traditional ties with the soil. In a commercial association the individuals involved have nothing in common other than their interest in a specific commercial aim. If that aim is once achieved then there is nothing left holding the group together. A community, by contrast, is held together by many bonds, which are likely to be deep and complex.

But any organization, however commercially orientated, is always liable to become more than a mere association of persons for a specific objective. The members will tend to have a single meeting-place, to adjourn for coffee after their discussions, to have an annual party, to have a small library, to give their surplus funds to a certain charity, etc. All these additional interests are liable to be added almost imperceptibly to the basic commercial objective, and they all constitute growing points for a *Gemeinschaft*. Indeed, far-sighted firms often encourage their workers to develop these

* F. Tönnies, *Gemeinschaft und Gesellshaft* (Leipzig, 1887, English translation by C. P. Loomis, *Community and Association*, London, 1956).

other ties by providing sports and other recreational facilities, by explaining or showing films about the products of the enterprise, and so on. Again, traditional industries, such as coal-mining, ship-building, or agriculture may be a focus for complete ways of life which dominate the basic attitudes of sections of the population. And the disruption of this, such as the coal-mining or railway industries have experienced in Britain, can produce a deep distress which is not adequately explained in terms simply of the loss of earnings.

The breakdown of a sense of community can be deleterious to the personality of the people involved. They become 'alienated' from the society to which they nominally belong. Marx sees the alienation of the worker from his work as an inevitable consequence of the exploitation which is built into the capitalist form of economy. Now we need not consider whether he is correct in his claim that alienation is the inevitable consequence of the capitalist form of economy, but we can make use of his concept of alienation to identify one aspect of the real need people feel to be emotionally involved in their community and its social structures.

It is in this context that one can introduce the idea of a 'culture'. The term 'culture' is often abused. For example, it is often used to refer simply to the arts and similar activities. And it is in protest against this use that some writers have said that science also is a 'culture', so that there are said to be 'two cultures' which are alleged not to understand each other. The truth is rather that when it is plausible to speak of 'two cultures' in this way it would be more accurate to speak of the breakdown of culture. A culture in the proper sense (that of the anthropologist) is a way of life in which the total activities of a society, including its arts and sciences and religion, are integrated and create for the participants a life of emotional significance as well as rational organization. It is the presence of the emotional bonds that can be indicated by means of the concept of fraternity; and any adequate account of social morality must make room for such a concept if it is to be comprehensive in its coverage of the facts of social life.

6. SOCIAL PRINCIPLES AND 'RESPECT FOR PERSONS'

In the account of social morality so far we have seen how people will tend to accept a form of social organization governed by the three principles of utility, equality, and liberty, and that a sub-rational cohesion is added by fraternity, understood as emotional bonds from a variety of sources. In stressing these four factors we have been guided by the criterion of comprehensiveness; the aim has been to provide an account which could reflect the main values of social morality. There will be still other, and important, concepts to introduce before the account can pretend to be at all adequate, but we shall pause in this chapter and attempt to streamline the account, as the criterion of economy requires. Is there any way of reducing the four factors to a single principle?

There can be no guarantee in advance that this will be possible. It may well be that social morality will defeat our hopes of a tidy scheme, and certainly the temptation to distort the facts of social morality, or omit them altogether, must be resisted. There is a story of someone who designed a shaving machine which took the form of a mask placed over the face. A button was to be pressed, blades would revolve, and the man would be shaved. Someone objected that not all faces were the same shape. The inventor replied, 'They soon will be'. There is always a temptation for a philosopher to take the idea of Ockham's razor rather too literally and to shave off inconvenient facts in the interests of a neat exterior. But I shall try to avoid this danger, and shall suggest that the ultimate principle of which the four so far discussed are specific expressions is that of respect for persons as ends (to use Kant's formula) or respect for the individual (to use the liberal formula). Let us consider how we might show the connection between the factors so far discussed and the principle of respect for persons as ends. Take first the principle of liberty.

It is clear that we stress the importance of liberty. The most plausible explanation of this stress is that we all recognize that freedom from external compulsion is necessary over a certain area of a person's life if he is to develop those attributes which make him characteristically a person. If a certain minimum freedom of action is violated by external compulsion then the individual will fail to achieve the level of self-realization which makes him a person: at first failing to express himself in characteristically human actions he will in the end fail even to envisage these actions as possibilities in his imagination. The principle of liberty is therefore the principle of respect for persons expressed in a context where the importance of self-realization is being weighed against that of achieving a harmonious and co-operative society. So, too, the principle of equality serves as a reminder in a social context of competing claims that *each* person starts off with the same claims to consideration as every other person. We may say that the principle of equality expresses that of respect for persons in a context of distributive justice. It is in virtue of the fact that they are all equally persons that no one individual has a better right to consideration than another, unless there are good reasons for unequal treatment. The principle of utility, when modified by those of liberty and equality, can be regarded as an attempt to create a social organization in which human beings can develop and express their personalities. Utility can be regarded in fact as an administrative expression of respect for persons, for it is unintelligible to say that we ought to maximize people's interests unless we presuppose the worth of persons in the first place.

Does the concept of fraternity also presuppose the principle of respect for persons as ends? If the feelings of kinship extend only to a limited group, a clan, a social class, or a group of football supporters, then it is not the same as respect, and may indeed be inimical to it. In so far as it tends to confine personal involvement to identification with some clearly restricted group, fraternity will be a divisive force in society. But where fraternity can be expressed by a saying like 'The whole world is my kith and kin', then it will presuppose and give emotional colouring to the principle of respect for persons. The relevant sense of fraternity may be expressed in other language, as by Hume, when he speaks of a 'feeling of humanity'. This is a sympathy which one person will have for another simply in virtue of the fact that both are persons. Sympathy of this kind is a part of the make-up of a normal human being, and its extension to persons as such makes sense only if we presuppose their supreme and equal importance.

The argument, then, is that our ordinary judgements of social morality presuppose the principles of utility, equality, liberty, and something which may variously be called fraternity, fellow-feeling or sympathy. These components can all be seen as expressions of the one principle of respect for persons as ends, which is Kant's way of stating the view that what is of ultimate value is the individual human being.

7. HUMAN RIGHTS AND 'RESPECT FOR PERSONS'

The argument can be put in a slightly different way by expressing it in terms of what have been called, at different historical periods, natural rights, the rights of man, and human rights. We have so far used the term 'rights' mainly in an institutional context, as when we discussed institutional rules, and this is a common use of the term 'rights' in the law or in everyday moral practice which we shall investigate further in Chapters 4 and 6 when we develop the idea of an institution. There are also rights of a more basic kind, however, and they belong to men as men and do not require the use of institutional concepts for their adequate identification or justification. Rights of this more basic kind are claimed and discussed in documents such as the American Declaration of Independence and the United Nations Declaration of Human Rights.

How can we identify human rights? It might at first be thought that we can identify them as the rights which appear on the characteristic lists of human rights, such as those we have just mentioned. Now it is true that most lists include the rights to life, liberty, equality, the pursuit of happiness, and property. The trouble, however, is that some lists also include other rights, such as those to a certain standard of living including holidays with pay, and, whatever is to be said for or against calling the latter 'moral rights', they do not seem to be rights in the same sense as the earlier ones. But by what criterion can we distinguish human rights from other moral rights?

The answer* to this question is that human rights are those which are *practicable*, *universal*, and *paramount*. Firstly, for something to be a human right it must be practicable for it to be granted. Clearly, however, it is impossible for all men to have a right to holidays, far less holidays with pay. Such a right may be practicable in prosperous societies, and could there be claimed by workers as a moral right, but since it is not practicable in all communities we cannot say that it is a human right or a right of man as such. Secondly, a human right must be something which *all* men can have. But since not all men are paid, not all men could have a right to holidays with pay even were this practicable. Human rights belong to men *as men* and not as wage-earners or the like. Finally, human rights are of paramount importance. Of course, it is true that the achievement of a certain standard of living or opportunities for education or employment are very important to people; but they do not have the paramount importance of liberty or equality. Let us say, then, that human rights are those moral rights which satisfy the tests of practicability, universality, and paramountcy. Can we further characterize them?

There is an important distinction between rights of two kinds: rights of action or liberties and rights of recipience.† It is characteristic of rights of action that they are asserted only when the exercise of them is in some way criticized. For example, if a young person is criticized for having his

* An answer for which I am indebted to Maurice Cranston, 'Human Rights, Real and Supposed', in D. D. Raphael (editor), *Political Theory and the Rights of Man* (London: MacMillan, 1967).

† This distinction is developed by D. D. Raphael, 'Human Rights, Old and New', in D. D. Raphael (editor), op. cit.

hair too long (or short) he might reply, 'I have a right to grow my hair whatever length I please.' The youth is here saying that it is not wrong that he should style his hair as he pleases; and this is the sense of a right of action or a liberty. To claim such a right is simply to deny that it is wrong to do whatever is in question, and to assert that it is *all right* or permissible to do it. A right of recipience, however, is a right that others should do or provide something. For example, if X has given his bicycle on loan to Y, then X has a right of recipience against Y to have his bicycle returned to him on demand. Y reciprocally has a duty to observe the right and return the bicycle.

Two points should be noted about this distinction. The first is that if X has a right of action to do action A – i.e., if it is permissible for him to do it – then X *also* has a right of recipience against people in general that they should refrain from interfering with the exercise of his right. It must be stressed that, while it is the case that whenever a person has a right of action he also has a right of recipience, these are two distinct rights. To bring this out consider the denial of a right of action. If we say to someone, 'You have no right (it is wrong) to treat your child so cruelly', our judgement is much more severe than the simple denial that there is a right of recipience that the person should be left alone. In other words, 'You have no right to' ≡ 'It is wrong to', and these are much stronger expressions than the denial of the right of recipience, 'No one has a duty not to interfere'. The latter may be implied by, but is clearly not so strong as, the former; but the two would be equivalent if there were no distinction between a right of action and a right of recipience.

The second point about the distinction which, like the first, emphasizes its reality, is that it does not follow from the fact that a person has a right of recipience that he also has a right of action. For example, as a result of an unfortunate bargain Shylock had a right of recipience to a pound of flesh, but it was not thought to follow that he had a right of action – that it was permissible for him to exercise the right.

My claim, then, is that whereas all who have a right of action in a certain area also have a right of recipience against interference, it is not the case that all who have a right of recipience also have a right of action.

Assuming that the distinction between rights of action and rights of recipience is roughly clear, I shall now assert that human rights are rights of recipience, rather than rights of action. It is true that human rights often concern liberty. But it is certainly false to suggest that all exercises of one's individual liberty are 'all right' or morally permissible. The human right of liberty is a right of recipience that one should not be interfered with in certain basic ways in the exercise of one's liberty, rather than the obviously false assertion that whatever one does with one's liberty is all right.

Against whom is a human right a right of recipience? The answer is that it is against mankind in general, and one's government in particular. This answer follows from the claim that human rights are universal; they are rights of all men against all men. But clearly in political society a man's government is particularly relevant since it pre-eminently has the power to protect human rights and safeguard the liberty of all men within its jurisdiction.

Now, if we agree that human rights are universal, practicable, and paramount moral rights of recipience, we can certainly see the rationale behind the restriction of such rights to a list including only life, liberty, equality, the pursuit of happiness, and property. Let us proceed by considering whether the list could be further shortened. I shall argue that it can be reduced to only two – liberty and equality – and that all the others can be shown to be reducible to either or both of the rights of liberty or equality. (To argue this, of course, is not to suggest that the traditional lists should contain only these two, for factors other than conceptual economy may rightly be operative in determining membership of the traditional lists.)

Take first the right to property. It might seem that the right to property is not in fact a basic human right at all but is like the alleged right to holidays with pay – a right which may be granted in some forms of social organization (capitalist ones, say) but not necessarily in all. This objection must be conceded if the right to property is taken to involve the right to own large estates, coal-mines, or railway networks. But when men have asserted that there is a basic right to property it may be that all that is meant (certainly it is all that it is plausible to claim as a human right) is that there is a right to own certain personal possessions through which a person can express his personality or individuality. After all, not even the most idealistic of communists, not even Plato himself in the *Republic*, ever argued that a man should not be free to possess his own personal toothbrush! In other words, we might argue that where it is plausible to claim a human right to property it will be possible to see this right as an expression of the right to liberty.

A somewhat similar analysis can be used to show that the right to the pursuit of happiness is simply an assertion of the right of liberty. For the only plausible sense in which we might all be said to have a right against all to happiness is that in which we ought to be allowed to follow our own lives as we choose without arbitrary interference. But to assert this is to assert the right of liberty.

The right of life seems to be a combination of the rights of liberty and of equality. For consider, life is so important to a man that almost no benefit to others can outweigh one man's loss of life. But this is really the limiting case of the minimum demand for equality or, to put it another way, the prohibition of extreme inequalities. We have discussed this already to some extent in the case of the 'death-mission', but let us take another example. Suppose there is a dangerous lunatic on the loose. It is obvious that he must be captured and removed from society, and the most economical way of doing this may be to dispose of him quickly and painlessly. Yet we should regard it as wrong to bring about the best possible consequences in this way for it would involve a violation of the minimum expression of equality – the equality of consideration. We can say, then, that 'Thou shalt not kill' is not simply an empirical generalization about actions of maximum utility but is an expression of minimum equality.

But this is not the whole story about the right to life, for what we want to preserve is not life in the minimum biological sense, but in the sense in which it is characteristically human. In other words, a right to life is a right to a life with a certain distinctively human *quality*. And the quality comes from the exercise of the right of liberty. It is in virtue of his ability to exercise his liberty in self-determination that the life of a human animal becomes distinctively personal life. And it is personal life which is at stake in the claim that we have a right to life.

My claim, then, is that basic human rights reduce in the end to the universal, practicable, and paramount rights of recipience of all men against all men, namely the rights of liberty and equality. Now, if these are the rights which are morally prior to all institutional rights and which restrict the morally permissible limits of institutional interference, they must be themselves intimately related to the supreme value of the individual person. But this can easily be seen. We have already argued that liberty is valued because it is in the exercise of his liberty that a person can develop those qualities which bring out his true nature. Thus the human right of liberty presupposes the supreme worth of the individual person. Equality too is an expression of the principle of respect for persons in a context of distributive justice. It is in virtue of the fact that persons are alike in being ends in themselves that they are entitled to minimum equality of consideration.

8. CONCLUSION

A survey of obvious features of human beings and their situation suggested that a certain very general form of social organization must be adopted if human beings are to survive in a characteristically human way. Utility seemed at first to be the principle governing that organization. (The claims of rational egoism, another prima facie plausible basic principle, are most persuasive in the sphere of motivation rather than organization.) Despite the immediate plausibility of utility, however, room must be made for other basic principles of organization, and in particular for the principles of equality, liberty, and fraternity. All four principles govern the social organization which, at least in Western civilization, is thought most suited to remedying the deficiencies of human nature. Using the criterion of economy we were able to 'streamline' the four principles and claim that they are all in different ways expressions in varying contexts of the one idea that the individual person is of supreme value. Moreover, if we approach social morality from the slightly different point of view of the language of natural or human rights, we can reach the same conclusion: that what matters supremely is the individual person.

3

Responsibility

1. THE MEANING OF 'RESPONSIBILITY'

The argument so far has been that in view of the basic features of human beings and their situation they will tend to accept and develop certain forms of social organization, in which the principles of utility, equality, liberty, and fraternity will have an essential part to play, and that if we consider what is presupposed by these principles we can see that it is the supreme principle of social morality – respect for the individual, or respect for persons as ends. In this chapter we shall discuss the sense in which the individual is responsible for his actions under these principles, and in what circumstances he is excused from responsibility. Let us begin by considering some senses of responsibility.

In the first place, we can correctly speak of one person as being *responsible to* another person or group. In this sense of 'responsibility' an employee is responsible to his employer, and the Queen's ministers are responsible to Parliament. 'Responsible' here means the same as 'accountable', where 'accountable' means 'being obliged to explain and justify what has been done'. In the second place, we can speak of a person being *responsible for* something, in the sense that it is his task or job or role to deal with it. This usage is often combined with the first, as when one says that a gardener is responsible *to* his employer *for* the proper upkeep of his garden, or that the Minister of Housing is responsible *to* Parliament *for* the state of the nation's housing. But we sometimes speak of a man's being responsible *for* something when there is no one to whom he may be said to be responsible. For example, we might say that an adult is responsible for looking after his own health. What a man is responsible for in this sense may be called his *responsibilities*. For example, we say, 'The care of your health is your own responsibility' or 'Your responsibilities as my gardener are to keep the garden looking attractive and to provide playing space for the children'. Thirdly, we speak of someone as being *responsible* and mean that he is reliable or conscientious or has 'a sense of responsibility'. Fourthly, a man can be responsible for something in the sense that he causes it to happen. This usage subdivides, however, into (4a) simply causing something to happen, and (4b) causing something to happen where there can be implications of praise and blame. For example, if I am pushed and fall through a shop window I am responsible for the breakage simply in being a causal factor. In this sense, a blocked carburettor can be responsible for the breakdown, or poor weather for the bad harvest. But we also say, 'He was responsible for the muddle in the arrangements' and mean not only that he muddled them but also that he was blameworthy; or we say, 'She was responsible for the beautiful floral arrangements' and mean that she arranged the flowers in a praiseworthy manner. Notice that this sense of 'responsible for' is different from sense (2) above. For, using sense (2), we might say, 'You are responsible for the flower arrangements and yet have done nothing about them!' For the purposes of this chapter the most important sense is that involved when we speak, fifthly, of 'the age of responsibility' or say that 'responsibility was impaired'. To be responsible in this sense is to have the ability to make up one's own mind rationally on what to do, and to be free to do it. All the other senses, except (4a), presuppose that the person concerned is already responsible in the fifth sense.

In order to understand more fully what is involved in attributing responsibility in the fifth sense to a person let us consider the circumstances in which we would say that a person is not responsible, or not fully or directly responsible, for his actions. These conditions – they were first examined by Aristotle* – are nowadays called 'excusing conditions' and they are reflected in our ordinary judgements of moral responsibility, and in a more formal way are operative in courts of law.

2. EXCUSING CONDITIONS

In the first place, we do not hold a person morally responsible if it can be shown that, while there is a sense in which it was he who performed the action, he was in some way *compelled* to do so. There are a range of situations in which this may be the case. For example, we do not hold a person morally responsible for knocking an old lady off the pavement if he can satisfy us that he could not help doing so because he was himself pushed. Sometimes, of course, such an excuse is not sufficient. We can imagine schoolboys indulging in horseplay as they walk along the pavement, and as a result of mutual jostling one of them is pushed into the old lady and knocks her off the pavement. In such a case we should absolve the boy from the charge that he acted intentionally, but should perhaps still hold him responsible in that he permitted himself to be pushed. As a rule, however, the excuse, 'I was pushed' is sufficient to absolve a person from moral responsibility in so far as it is taken to establish that he could not help doing what he did. And in general the presence of external physical compulsion is incompatible with the attribution of moral responsibility.

There is a second sort of physical compulsion which differs from the first in that it is 'internal' to the agent rather than 'external'. One kind of example of this is the movements of the body which may occur as the result of a disease in a person's nervous system. Such movements are like the ones which take place when a person is pushed, but their causation is 'internal' in that it is the outcome of a disorder in the person's own body. For example, someone suffering from *chorea* (St Vitus' dance) might knock over a vase as a result of a convulsive muscular movement. In such a case we should

* Aristotle, *Nicomachean Ethics*, Book III, Chapter 1.

not hold him morally responsible for what he did, although in certain cases we might blame him for putting himself in a position where accidents were liable to occur.

A third sort of compulsion should be regarded as psychological rather than as physical. Like the second sort, from which in some cases it will not easily be distinguished, the compulsion will be 'internal', but the causation will not in standard cases have a clearly physical basis. For example, it is sometimes claimed that theft of certain kinds – shoplifting, for instance – can be a crime towards which some people are irresistibly pushed by a quirk in their minds. There are clearly practical problems for psychiatrists and juries in deciding when cases are genuinely instances of irresistible impulses and when they are simply instances of impulses which have not been resisted. But if it can be established that a person was compelled by some internal psychological force to act as he would not have chosen to do, or as a normal person would not have chosen to do, we absolve him from moral responsibility for his actions.

A fourth sort of compulsion raises rather different issues. It is illustrated in the actions of a man who is threatened with violence unless he does what he is told to do. For instance, a person may be forced at gunpoint to sign a document he would not otherwise have signed. A different kind of example of the same sort of compulsion is provided by Aristotle's account of the sailors who must either throw their cargo overboard or be shipwrecked. Compulsion of this sort is often called 'duress'.

Now sometimes it is thought that actions performed as a result of duress are like the actions performed as a result of compulsion of the other three sorts, in that they do not carry with them moral responsibility. But this seems too hasty a conclusion. For it may be said that even if we would not call an action performed at gunpoint *voluntary* – since it is certainly not what we want to do – there is still a sense in which it is *intentional*, and as such must carry moral responsibility. If we do not blame a man for yielding to duress, it is surely because we think that what he does is *in the circumstances* the best thing to do. Thus the bank-clerk who is threatened at gunpoint may be praised as prudent, sensible, etc. if he hands over the money, and blamed as reckless, stupid, etc. if he does not, and Aristotle's sea-captain would be blamed as avaricious if he were to risk the crew's life rather than jettison the cargo. On the other hand, as Aristotle points out, there are actions of such a kind that a man should be prepared to die rather than perform them. Here the difference is that what the man is 'compelled' to do is so dreadful that no threat could make it the lesser of two evils, although we would regard the duress as a mitigating circumstance, rendering the deed less dreadful than in other circumstances it would be. Extreme fear or suffering may make a man incapable of *choice*, of course, and here we would have a situation in which we could no longer speak of responsibility because one of its necessary conditions would be absent. But in general the kind of compulsion called 'duress' does not absolve a man from moral responsibility for what he has done, however it may *mitigate* his blameworthiness. Duress differs, then, from the first three types of compulsion which do absolve from moral responsibility. Let us now consider 'excusing conditions' of another kind.

In general, a successful plea of *lack of knowledge* is sufficient to absolve a person from moral responsibility for specific actions. As in the case of compulsion, there are several subdivisions of this plea. In the first place, a person may not know all the relevant facts of the situation in which his action has been performed. He acts in ignorance and his ignorance removes his responsibility for the actions in question. This is not to say that it removes his responsibility for other actions or failures to act in the past. For instance, perhaps he ought to be in possession of the relevant facts and his failure here is a sign of culpable laziness in the past. This can all be admitted – indeed, it must be insisted upon – but it does not affect the contention that if we can show that a person has acted in ignorance of the facts he is not to be held morally responsible for the action which was done in ignorance. But notice that ignorance absolves a man from responsibility concerning only that aspect of the action about which he was ignorant. Thus Oedipus, though not responsible for parricide, was responsible for murder, because he did know he had killed an old man even although he did not know the old man was his father. 'The action which was done in ignorance' is really the action under one particular description.

A second form of lack of knowledge is more naturally described as acting by mistake. For example, a man may know the facts that his coat is black and that he left it near the end of the row of pegs, and yet in the dim light of the cloakroom he may by mistake take the wrong coat. In so far as it seems plausible to believe that what was done was done by mistake it will seem plausible to waive the imputation of moral responsibility. Once again, of course, there is room for qualifications, for in certain cases a person can be blamed for making mistakes. These are mainly cases in which we think that if the person had taken the care before acting which we expect a reasonable man to take he would have been in a position to know better what he was doing and would not have made a mistake. His lack of knowledge in this respect is therefore a matter of negligence and has moral responsibility attached to it. There are cases, however, in which even a reasonable man might not have avoided making a mistake. In these cases there is no negligence and moral responsibility is waived.

So far we have distinguished two (closely connected and overlapping) kinds of lack of knowledge – acting in ignorance of the facts and mistaking the facts – but there is a third kind of lack of knowledge where, although the facts are known and not mistaken, their moral significance is not known. The law recognizes such cases, in their pathological form at least. Thus, according to the M'Naghten Rules, a man should not be held responsible for his actions if it can be established that at the time of acting he was suffering from a defect of reason such that either he did not know the nature and quality of his action (did not know what he was doing), or, if he did, did not know that it was wrong.

In the M'Naghten Rules 'wrong' means 'illegal', but sometimes a person may not know that actions of a certain kind are morally wrong. There are, of course, the pathological cases of this: moral insanity occurs as well as the more familiar kinds. But more interesting are cases which are not in general pathological, but which may seem appropriate for being classified as 'moral ignorance'. The ignorance may be due to the bad choices in the past and as such be culpable, but it may also be due to bad upbringing or an unimaginative or underdeveloped nature. In such cases we absolve a person from moral responsibility; we absolve him because, as we say, 'He does not know any better'.

Ignorance of moral matters may be shown in a slightly different way. There are cases in which a person may know that actions of a certain kind are wrong but not realize that what he is doing comes under the general rule. It may be said that such cases – cases of ignorance of the minor premise or the premise that states that a particular action is governed by a certain rule – are simply forms of the mistake of fact already discussed. This will sometimes be so, but there are also occasions in which a failure to know that an action is governed by a rule is a case of moral ignorance; for there is not a clear distinction between deciding what the relevant facts are in a given case and making a moral assessment of them. And moral ignorance may be illustrated in a failure to see the significance of facts. For example, a man may fail to see his false Income Tax return as a case of dishonesty. Moral ignorance may therefore be shown equally in lack of knowledge of the major premise – that actions of a certain type are wrong – or in lack of knowledge of the minor premise – that this action comes under an admitted rule.

'Excusing conditions' can be said, then, to be of two main types: cases of various kinds in which a person is compelled in some sense to act as he would not otherwise have done, and cases in which a person in some sense lacks knowledge of what he is doing. It may be said that if we accept this answer to the negative question about when we withdraw our assumption that persons are morally responsible for their actions, we also confirm the positive conditions for moral responsibility already mentioned (p. 56). And certainly in a sense there can be little objection to saying that a person is morally responsible for his actions if he performs them freely and with full knowledge of what he is doing. But it must be stressed that these are not criteria by which we establish positively moral responsibility for actions. The force of the tests is a negative one; we require to be persuaded out of our belief that people are morally responsible for their actions.

3. DETERMINISM

It is possible to devise arguments, however, which, if valid, will entail that the concept of responsibility as usually understood has no real application. These arguments are not concerned with the second of the grounds on which a person may be absolved from moral responsibility in specific cases, for it has not seriously been argued that people in general are not morally responsible because they never really know what they are doing. But there have been many theories to the effect that people in general are not to be held morally responsible because their actions are never really free. In other words, the claim, familiar to all, that on certain occasions we are not responsible for our actions because we may have been compelled in some way to act as we did, is generalized to apply to all our actions. The result is a doctrine which is far from familiar to our ordinary thinking. If valid it will mean that our ordinary way of regarding our own actions and those of other persons is radically mistaken; our belief that we are morally responsible for our actions will be either completely erroneous or else in need of radical revision.

Yet the considerations which may tempt us to argue for this paradoxical conclusion are themselves familiar enough. They can be described, grandly, as the considerations which make science a rational pursuit, or, more humbly, as the considerations which enable us to make sense of the events in our everyday life. They are, in short, the assumptions implicit in the spectator attitude to the world. We all can and do regard the world not only from the standpoint of agents with intentions but also from that of spectators of its events. Now essential to the spectator attitude to the world is the principle that every event has a cause. It is essential to the extent that it is not possible to make sense of the interactions of things in the world without assuming the principle. To 'make sense' of the world of things and events is in fact to find causal explanations which assume the existence of causal laws. It may be objected that science no longer assumes the universal validity of the principle of causality because a branch of science – quantum mechanics – proceeds without assuming the principle, and indeed suggests that the principle does not hold for the submicroscopic world it investigates. The laws of quantum mechanics, it may be said, are not causal but statistical. However that may be, we have been given no reason to suppose that the principle of causality does not govern the macroscopic world of the senses, and it is with the world of the large scale that the majority of the special sciences are concerned. Let us assume, then, that the principle of causality governs the scientific investigation of the large-scale world as well as our ordinary spectator outlook. How does this assumption lead us to doubt our freedom as agents, and consequently our moral responsibility?

The answer to this question emerges in outline if to the thesis that every event has a cause we add another, that human actions are like events in many respects. This may lead us to consider plausible the hypothesis that human actions have causes. Now if we interpret a 'cause' – as it is natural to do – as 'that which makes an event happen', we seem to be forced to interpret our conclusion in a way which implies that our actions are 'made to happen' and so are not free. But we have already seen that if actions are not free we feel justified in withdrawing the assumption that they are morally responsible. Hence, starting from the spectator viewpoint of the world and proceeding by steps which seem easy to take, we have finished with a conclusion which contradicts one of our basic assumptions about persons. To accept the validity of such a pattern of argument is to accept the thesis of determinism.

Philosophers have reacted in different ways to determinism. There have been some who have taken a tough-minded line and maintained that since determinism is the thesis which seems to express the basic assumption of science, and since science is clearly a valid enterprise, determinism must be correct and hence freedom must be an illusion, and likewise illusory must be our ordinary conviction that as agents we really do act and are responsible for our actions. This line is difficult to accept, however, because it requires us to discard as erroneous one of our basic convictions about human beings. Another tough-minded line is to say that because we are convinced of our responsibility as agents, and because such responsibility implies freedom from causal necessity, then determinism cannot apply to human actions. But some philosophers might say of this that it smacks of the refusal to look through the telescope. It is a fact, they might say, that psychologists and others investigate human actions as if they were events and that they are making some headway in providing causal explanations of them and in predicting the conditions in which they are likely to occur.

One or other of the tough-minded lines may well be the plausible one to take in the end, despite the initial difficulties.* But many philosophers prefer to argue that the contradiction between the agent-assumption of responsibility and the spectator-assumption of causal necessity is only apparent. By stating in more detail what the assumptions of the agent and the spectator really amount to they hope to show that there is no contradiction between them. Let us consider one such attempt at reconciliation. We shall call this the 'reducibility thesis' because it attempts to reduce the language of choice to that of causality.

According to the reducibility thesis, an action is produced by the circumstances in which an agent must act and the

* The tough-minded Libertarian line is taken in Downie and Telfer, op. cit., Chapter 4.

agent's desires at the time of action as they arise out of his conception of the circumstances and his long- or short-term aims. The claim is that the circumstances and the agent's desires jointly make up the causal conditions from which the action follows.

One merit of this formulation of determinism is that it seems to avoid assimilating action to process, because one of the essential causal conditions has been identified as the agent's desire to do whatever it is. Indeed, we can still speak of 'free action' on this account, for, if the agent's desires change, the action will correspondingly be different. Thus, something like our ordinary conceptions of choice and action remain, but on the other hand the assumption of universal causality is also maintained since a desire is said to be a cause. In other words, the relevant causation is regarded not as something external to the agent which 'compels' him against his will, but rather as his own will itself, interpreted as his desires at the moment of action.

Another merit of the analysis is that it enables us to explain certain familiar phenomena of everyday social life. In the first place, we regard it as important to train a child so that its desires are directed towards socially permissible or desirable ends. And in general moral education is directed at character-building and the creation of a socially desirable pattern of desires. Now it may be said that there is no point in such processes of moral and social education unless desires produce actions as causes produce effects. Thus it may be argued that we cannot account for the significance we give to character-training unless we assume the thesis of psychological determinism that desires are the causes of actions. In the second place, we all do make inferences from actions to character. We do not regard people's actions as random events but rather as proceeding from a character of a certain sort. Just as a tutor may regard a student's work over a certain period as a sign of whether he has a good mind or a moderately good one, so we all do regard people's actions as signs of whether they are generous, courageous, courteous, and so on. The determinist argues that the fact that we can make such inferences speaks for the truth of determinism, for such inferences could not be well-founded if actions did not proceed from desires as causes, since character is only the pattern of one's desires.

Despite these merits the theory of psychological determinism has some serious disadvantages which make it in the end very difficult to accept. Consider first of all the implications of psychological determinism for our ordinary activities of praising and blaming, rewarding and punishing.

The ordinary view of praising and blaming is that we praise or blame an agent for an action he has already performed. Similarly, and perhaps even more obviously, in ordinary life we regard punishment and reward as being for what a person has done. On the determinist scheme, however, it would seem to be pointless to praise or blame, reward or punish for what a person has done, for at the moment of choice the agent had no control over his desires and so could not have done otherwise. There therefore seems to be an incompatibility between the implications of determinism for praising and blaming, etc. and our ordinary assumptions about these activities.

Determinists would say in reply to this argument that the incompatibility is only an apparent one, for they would not usually draw from the premise, that the agent at the moment of choice could not but act as he did, the conclusion that it is pointless to praise or blame, reward or punish. Rather, they would provide a new significance for praising or blaming, rewarding or punishing. Praising and rewarding, blaming and punishing, determinists might say, are activities in which people engage. The aim of the activities of praising and rewarding is to encourage people to persist in types of action which have desirable results. Likewise we blame and punish to discourage people from persisting in actions which have undesirable results. Where the activities of praising and blaming, etc. will not have these good results we ought not to engage in them.

The determinist account is an ingenious one, but it is not supported by our ordinary conceptions of praising and blaming, rewarding and punishing. In our ordinary thinking there is a distinction, which is no less basic for not always being made explicit, between saying that someone is blameworthy with respect to a certain action and saying that it is right to blame him for it. There are in fact two different issues involved here and not just one, as the determinist implies. A person may have performed an action for which he thoroughly deserves blame (that is, he is blameworthy for it) and yet it may not be right to blame him for it. This point emerges more clearly when it is pointed out that 'to blame' means in this context 'to reprove'. Suppose, for example, the person is about to sit an important examination and would be upset by being blamed in the sense of reproved, or is about to have a baby and would be likely to have a miscarriage if punished. Yet such persons would remain blameworthy. It is in its inability to provide a satisfactory account of how a person can be blameworthy (or praiseworthy) that the defect of determinism is illustrated. Determinism can provide an account of how blaming and praising can exercise a causal stimulus which will affect a person's future behaviour. But we can still ask how praising and blaming are to be justified, and a necessary (although perhaps not sufficient) condition of the legitimate exercise of these activities is that the agent be blameworthy or praiseworthy. But how can an action be blameworthy or praiseworthy if at the moment of choice the agent had no control over what he was doing?

It is true that we may praise or blame a child or an animal for certain actions, and that this may be aimed entirely at affecting the future behaviour of the child or animal. But this is surely different from praising or blaming the mature person. The point is that praising and blaming, rewarding and punishing are not just forward-looking; they also look backwards to the action which has been performed, and the difficulty for the determinist is in understanding how, from his position, the backward-looking aspect of these concepts can be accommodated.

The inability of psychological determinism to reflect our ordinary conceptions of praising and blaming, rewarding and punishing, is a specific form of the second main difficulty

I shall raise. The second difficulty concerns the account the psychological determinist gives of 'acting otherwise'. The statement, 'He could have acted otherwise', must always be translated by the psychological determinist as 'He would have acted otherwise *if*...', where 'if' refers to a hypothetical desire, 'if he had wanted', or to hypothetical circumstances, 'if things had been different'. Now these conditions may not be fulfilled: perhaps the agent did not want to, or perhaps things were not different. But if one or other or both of these causal conditions did not obtain, then the agent could *not* in fact have acted otherwise. This view of action, and of the agent's control over it, does not seem to justify our ordinary conception of moral responsibility, for that conception requires that absolutely he could have acted otherwise. The same point could be stated in a slightly different way if we said that on the determinist scheme, 'He could have acted otherwise', must be translated as a subjunctive conditional, 'He would have if...', whereas our ordinary view of responsible action requires that 'He could have . . .' be taken grammatically at its face-value – namely, as a past indicative: 'He could have', full stop.*

The third difficulty I shall raise for psychological determinism is the most radical, for it concerns the basis of the whole theory – the analysis of choice or decision in terms of desire construed as the cause of action. If our actions proceed from desires as causes then it would seem to follow that our knowledge of what we are going to do will be obtained in the way in which we obtain knowledge of causes – by experience or inductive procedures. But our knowledge of what we are going to do does not seem to be like this at all. We do not say to ourselves, 'Experience teaches that when I have a certain desire I always act in a certain way. Hence, I know that I am going to read a book.' If I know that I am going to read a book it is because I have decided to do so. We may say, 'Experience teaches that when I eat porridge late at night I have indigestion.' But this is causal knowledge of what is

* cf. J. L. Austin, 'Ifs and Cans', *Proceedings of the British Academy* (1956), collected in *Philosophical Papers* (Oxford: Clarendon Press, 1961).

going to happen, not of what a person is *going to do*. The psychological determinist may therefore be confronted with a dilemma: either desires are not causes (or not like other causes) or choice cannot be analysed in terms of desire. In either case the basis of his theory is undermined.

The conclusion of this discussion is that psychological determinism must be rejected as an account of our ordinary views of moral responsibility. It should be pointed out that there is a great deal more to be said either way on this question. There are other forms of determinism as well as the psychological thesis with which we have mainly been concerned here, and there are levels of analysis other than that at which it is appropriate to operate in an essay on social ethics. The point is that if psychological determinism were true it would have implications which are at variance with ordinary views on social morality (or so I have argued). It may be that there are other ways of stating determinism, or other levels of analysis where it is less appropriate to test the validity of a theory in terms of its ability to accommodate the facts of social morality. Since we are not concerned with metaphysics, and there is a great deal more to be said on other topics within social ethics, I shall confine myself to sketching an outline of a positive theory which approaches more nearly ordinary assumptions about responsibility within the context of social morality.

4. A POSITIVE THEORY

The postulate of the spectator standpoint has been formulated as 'Every event has a cause'. But the term 'cause' is ambiguous: it can mean that which is a sufficient condition of an event, or that which is a necessary condition. As a sufficient condition a cause is that on the occurrence of which an event necessarily follows, but as a necessary condition a cause is that without which an event cannot take place. Now the arguments against determinism are really only arguments against the view that actions have causes as sufficient conditions. It is quite possible to retain the ordinary conception of social morality while conceding that actions have causes as necessary conditions. Indeed, one must insist on this. To allow that actions have causes as necessary conditions is to make a move with two merits: it enables us to grant to psychologists and others the right (which they assume anyway) to investigate the causal antecedents of action, and it enables us to make room for the operation of reasons, purposes, intentions, and (what is fundamental) *choices* as causally irreducible factors in the explanation of responsible action. This solution requires expansion, and it may be helpful to expand it by answering three questions: Is it legitimate to speak of a necessary condition as a cause? What are the necessary conditions of action? Is it possible in any cases to provide a sufficient causal explanation of action?

It does seem to be legitimate to use the concept of cause for what is only a necessary condition of action. Take for instance the case of an accident. From one point of view – that of the police, say – the cause of the accident was drunken driving. But from another – that of the garage mechanic, say – the cause was a defect in the braking system of the motor-car. From a third point of view – that of the road engineer, say – the cause was a bad camber at a bend in the road. All three factors (and perhaps others as well) can equally be regarded as necessary conditions of the accident, and in a given case no one may be sufficient to cause the accident. But equally each may be regarded as 'the cause' from a restricted point of view. This seems a natural and not uncommon usage of 'cause'. And when scientists investigate the causes of action they may be investigating only necessary conditions; and they single out as 'the cause' of action those necessary features of the action in which they are interested from the point of view of their science – psychology, neurology, or the like. This is legitimate, but it can become confusing if their activities are thought to be directed at the discovery of sufficient causal conditions of action.

In answer to the second question we can say to begin with that the necessary conditions of action are of two kinds:

causal and non-causal. The necessary causal conditions of action are various. Suppose we take the action of posting a letter. The necessary causal conditions of this will include the movements of muscles and bones, the transmission of nervous impulses to the brain, and so on. Now without the occurrence of these complex sets of events the posting of the letter would not take place. But to list all these necessary causal conditions is not to provide a sufficient causal explanation of the action. To do this we need to bring into the explanation factors such as the agent's *beliefs* that this is a pillar-box, that he is in time for the last collection, etc., his *reasons* for wanting to catch the last collection, and his *decision* to do so. These latter conditions are also necessary for an adequate explanation of the action of posting a letter, but they belong to a different logical category from the former in that they are not reducible to causal terms. Thus, the necessary conditions for the occurrence of a normal action are of two kinds, causal and non-causal, and both are necessary for a complete explanation of action, although for specific purposes we may concentrate on one kind or the other.

The third question to be answered is whether there are ever contexts in which we can regard causal explanations of action as sufficient.* In the first place, we can provide sufficient causal explanation of the failure of an action to attain its goal. For example, it is not possible to provide a sufficient causal explanation of why a man entered for a race, but it may be that there can be a sufficient causal explanation of why he did not win it, namely that he slipped as he started, or he was not in training. Again, if we can establish that a person has not got a good brain we have sufficient causal explanation of why he did not succeed in becoming a great mathematician. In the second place, sufficient causal explanations can be provided of things which seem like actions but are not really so, such as slips of the tongue, obsessional handwashing, etc. It was one of Freud's major contributions to suggest causal explanations

* I owe some of the arguments here to R. S. Peters, *The Concept of Motivation* (London: Routledge & Kegan Paul, 1958).

for such goings on. But the explanations he provided bring out that these matters are to be seen as things which happen to a man, rather than as things which he *does*. To provide causal explanations of the Freudian type is to deny that explanations in terms of reasons or choices are appropriate – and to deny this is to deny that what we are explaining is really action in the full sense. Hence, sufficient causal explanation of action is not in fact possible. An adequate explanation of action requires *two* irreducible types of necessary condition: causal conditions and the agent's rational choice to perform the action. Of course, in this context 'adequate' simply means 'able to do justice to ordinary conceptions of responsibility'.

5. CONCLUSION

In this chapter we considered the concept of moral responsibility, but confined the discussion, as far as possible, to those issues which are relevant for a general theory of social morality (as distinct from metaphysics). After considering the various senses in which the term 'responsibility' is used, we identified the basic sense for our purposes. We ordinarily assume that in this basic sense a man is in general responsible for what he does. There are, however, circumstances – 'excusing conditions' – in which we allow that a man is not responsible, or not fully responsible for his actions. The thesis of determinism – that actions, like other events, have causes as their sufficient conditions – seems at first to conflict with this assumption of responsibility. Determinists have tried to reconcile their views with ordinary convictions of responsibility by inviting us to regard the agent's choice as a causal condition. Determinism, however, has implications which are hard to square with ordinary convictions. For example, it forces us to reinterpret concepts such as praise, blame, reward, and punishment in a forward-looking direction, whereas they are essentially backward-looking concepts. Again, determinism entails an interpretation of 'He could have acted otherwise' which might be said to be at variance with an ordinary interpretation of the expression. Finally – and most fundamentally – the analysis of choice in terms of desire construed causally is open to the criticism that it is at variance with our knowledge of what we are going to do, for such knowledge is not based on induction or on experience but on our decision to act in a certain way.

Having rejected determinism I suggested very tentatively the outline of an alternative theory of the basis of moral responsibility. In terms of this theory actions have necessary conditions of two categorially irreducible sorts – causes and choices – and a logically complete explanation of action would require us to bring in both, although, of course, for everyday purposes we often concentrate on one or other type of condition and assume the operation of the other.

4

Authority, Legitimacy, and Representation

I. THE CONCEPT OF AUTHORITY

There is still an important range of phenomena which cannot be accounted for in terms of the theory of social morality so far developed. For example, a policeman can arrest a man, we are all obliged to make tax returns and to pay the taxes which a government may levy, houses can be compulsorily purchased, men may be obliged to serve in the armed forces, and so on. It is not clear how these phenomena can be fully explained using only the concepts we have so far discussed. We have indeed mentioned some of the concepts which are here important, such as that of institutional rules, but the concept of an institution requires to be further explained if we are to understand the type of social phenomena just mentioned. The concept needed for this purpose is that of authority. In analysing the concept of authority* I shall distinguish two questions which are easy to confuse: that of the nature of authority, and that of what makes authority legitimate.

Writers often begin an analysis of the concept of authority by contrasting it with power, where 'power' may be defined in general terms as 'the ability to bring something about'. The contrast is helpful in bringing out several points about authority. In the first place power can be (and characteristically is) coercive, whereas authority as such is never coercive. Thus a highwayman uses his coercive power to stop the coach and remove the valuables of the passengers, but it

* My analysis of authority owes much to R. S. Peters' and Peter Winch's Symposium in *Aristotelian Society*, Supp. Vol. 32 (1958), reprinted in Anthony Quinton (editor), *Political Philosophy* (London: O.U.P., 1967).

would be very odd to see this as the highwayman exercising his authority. It might be objected that a policeman arresting a violent criminal uses coercion. This may be true but it does not show that authority as such can be coercive; it shows only that where someone has authority to act or forbear in a certain way it may be legitimate for him to use coercive power, if he can. He may not have coercive power at his disposal, however, but we should still say that he had authority. For example, the policeman may be overpowered by the violent criminal, but it remains true that the policeman had authority to arrest him. Thus, in so far as power may involve, and authority as such never involves, coercion we have a point of difference between the concepts of power and authority.

The matter is not so simple, however, because power need not involve coercion at all. For example, a dominant parent may have power over a child, but it does not necessarily follow that the parent in any way coerces the child; the parent may simply cause the child to do certain things. Now it may be thought that, in so far as it does not involve coercion, this usage of 'power' is like a usage of 'authority'. But always in the concept of authority there is the suggestion that the exercise of authority is in some way legitimate or satisfies certain criteria, whereas this need not be so at all with power. One person may cause another to do various things and we might on that account say that he had power over the other; but we would not say that he had authority unless there was more to the relationship than the exercise of power or influence. Hence, 'authority' always carries with it some suggestion of legitimacy, whereas 'power' as such never does. This is true of all the uses of the term authority. Let us now consider these uses.

Firstly, we can speak of a person as being 'an authority' on something. To call him 'an authority' is to call him an expert, and to suggest that his pronouncements on his subject-matter are liable to be correct; he can therefore, secondly, speak 'with authority' on his chosen subject. Here the idea of legitimacy is again present; since the person knows what he is talking about he can legitimately make statements about his subject with the conviction of an authority, and expect to be believed. Thirdly, we speak of a person as 'having authority'. This is the sense of authority which, as we have seen, seemed similar to a sense of 'power', but the difference was that to have authority with someone is *legitimately* to cause him to act in certain ways. For example, in a crisis some people stand out and can be said to have authority. This means that we regard it as legitimate that they should order us around, as the admirable Crichton did in Barrie's play. Fourthly (and most importantly for present purposes) we speak of a person as being 'in authority'. This is quite different from having authority. Thus, the admirable Crichton may have authority on the island, but we would not say that he was in authority, any more than we would say that some person who organizes escapes in a fire with authority is in authority. Conversely, a person may be in authority but have no authority, i.e. be unable to exercise the power, coercive or causative, which would be legitimate. Thus, we might say that those in authority in a college ('the authorities') had no authority with the students. But, despite the differences between 'having authority'* and being 'in authority', there is the implication of legitimacy about both. Fifthly, we speak of one person authorizing another to do something. To say this is to say that one person can somehow make it legitimate for another to do something. If the first person is in authority he can perhaps delegate some of his responsibilities to another. Or it may simply be that one person makes certain arrangements for another legitimately to act on his behalf. In all cases of authority, then, there is necessarily some sort of legitimacy although there may or may not always in fact be accompanying power.

2. 'LEGITIMACY' AS A LEGAL CONCEPT

Let us now consider what makes authority legitimate. The answer to this question will obviously depend on the kind of

* Note that *having authority* is not the same as *having authority to. . . .* The *authority to* do something can be possessed only by someone who has been authorized. See the fifth sense.

authority under consideration. For example, to be 'an authority' on a given subject is to have come up to the standards of excellence recognized by practitioners of the subject. A rather different case is that sometimes called 'charismatic authority'. A charismatic authority has authority with people, in the sense that whatever he tells them to do or believe they will in fact do or believe. Some great religious or political leaders are said to have authority in so far as their followers believe them to be appointed by God to be the vehicle by which a national, racial, or moral goal will be realized. Those who disbelieve in the legitimacy of the ends towards which the leader directs his followers are more apt to say that he has charismatic *power*, thus admitting the fact that the leader can cause people to act in certain ways but denying the legitimacy of the ends by withholding the term 'authority'.

It is not so easy to see straight away what it is to be legitimately *in authority*, and that is the sense of authority which concerns us most. Let us return to the paradigm situation of the exercise of authority in the relevant sense. The policeman arrests a criminal, and we all concur in this social phenomenon. How is this to be accounted for? It is true that arresting a criminal may be required by the principle of utility, since the criminal is causing harm to society. But why is it the policeman who is especially appropriate here? Or, if it is objected that in some circumstances an ordinary citizen can arrest a criminal, we can change the example and ask why is it appropriate that the postman should deliver the letters to our doors. It may be said that it is appropriate that they should do these things because they have been authorized to do so. But who has authorized them to do so? The Chief Constable and the Postmaster-General? To reply in this way is to invite further questions about who authorizes the Chief Constable or Postmaster-General, and indeed how a person comes to be a Postmaster-General, a policeman, or a postman. Let us try to answer these questions, beginning with the question of how it is possible for a man to be a policeman, postman, or the like, or in general, how it is possible to be *in authority*.

The picture of social organization as so far drawn in our earlier chapters is essentially one in which there are rules that people tend to accept because they compensate for people's natural inadequacies and deficiencies. These rules can all be placed under the general headings of utility, equality, liberty, and fraternity, and we suggested that these headings were all expressions of different aspects of the idea of the supreme importance of the individual person. Now to explain how a person can be in authority we require to introduce the idea of a second set of rules which belong to a different logical order or category. These rules will be second-order in that they govern the first set in several different ways. Let us consider these ways.*

In the first place, second-order rules will state what is to count as a first-order rule. There is a need for such a second-order rule in that in any given community first-order rules of many kinds and with many different origins will abound and there must be some way of deciding which rules the community will recognize as officially binding. Hence, there is the need for a second-order *rule of recognition*. This will say that first-order rules issuing from certain sources – such as custom, the wills of contracting parties, the precedents of certain courts, and the State legislature – will be recognized as binding. There is no logical limit to the types of first-order rules which may be recognized – a second-order rule of recognition may recognize first-order rules from a large number of sources or from only one, such as the decrees of an arbitrary tyrant – but clearly all developed forms of society require some sort of second-order rule of recognition, even if it is only that whatever is written on a certain parchment or tablet of stone will count as a first-order rule.

In the second place, a second-order rule will lay down how and when first-order rules may be altered. Clearly, again, there is a need for such a second-order rule in any society. Societies change in respect of their basic beliefs, their

* The analysis of secondary rules is based on H. L. A. Hart, *The Concept of Law*, Chapters V–VI (Oxford, 1961).

technological potentialities, their relations with other communities, their resources, etc. Such changes may call for changes in the first-order organizational rules of the society, and every society must have some means of determining the appropriate changes. There is therefore a need for some second-order *rule of change*. This, again, may vary from the simple second-order rule: 'If the tyrant says the first-order rules have been changed, then they have been changed', to the more complex legal procedures of a modern government. But the principle in each case is the same.

Thirdly, in any society some people must be empowered to carry out the first-order rules or to enforce them. For example, if it is the Home Secretary who authorizes the policeman there must be second-order rules which empower the Home Secretary to authorize policemen, or (to put the same point in another way) there must be second-order rules which set up the office or role of Home Secretary. Once again these *empowering rules* may vary from the highly complex legal rules to the simple authorizing of a henchman by a tyrant, but the principle remains the same in each case.

Fourthly, in the drawing up or carrying out or changing of first-order rules there are commonly appropriate procedures. For example, if two parties make an agreement it is useful if certain procedures are recognized as appropriate. Hence, there is a need for second-order rules stating what are the appropriate procedures to be gone through in invoking or in any way making use of the first-order rules. Second-order *procedural rules* will vary greatly from those stating what a judge may or may not say to a jury to those stating that vacancies in posts of certain kinds must be advertised.

We have, then, second-order rules of four types (perhaps there may be others) and any adequate account of a social system must make room for them, for there are many phenomena of social life which cannot be explained without them. In particular, we can now understand how a person can be in authority. If his role has been constituted in terms of the appropriate second-order empowering rules, or if the person authorizing him has himself the authority to authorize in terms of the empowering rules, then the person can be said to be in authority.

But is his authority legitimate? It might be replied to this that a person's authority is necessarily legitimate if it is constituted by the relevant second-order empowering rules of his kind of society. Now this reply is all right as far as it goes, but it leaves unanswered an important question or set of questions. This set of questions may be exposed if we ask about the nature of the legitimacy which is involved. In other words, is there a distinction to be drawn between moral and legal legitimacy, and, if so, how is it to be drawn?

In answering the first part of the question we may safely say that there is a distinction to be drawn between legal and moral legitimacy. It is true that some medieval supporters of Natural Law doctrines did not admit such a distinction and for them whatever was morally repugnant could not count as legally binding. And the fact that legal and moral terminologies overlap to such a large extent still encourages confusion on this point. For example, we might say of a law which excites moral disapproval, 'That's no law at all'. Or again, the policeman who extorts a confession and the false witness are behaving in a morally wrong way, but their behaviour is also legally illegitimate; and it may be thought that it is because it is immoral that it is illegal. But it is illegal merely because it is not in accordance with second-order rules of procedure. Of course, as we shall see, there are connections between second-order rules of procedure and morality, as there are between first-order laws and morality, but the connection is not of the direct because-it-is-immoral-it-must-be-illegal kind. If we assume, then, that there is a valid distinction to be drawn between legal and moral validity it is not too difficult to say wherein consists legal validity. A rule is legally valid if it satisfies whatever happen to be the second-order rules of validity in the society in question.

Consider, for example, some of the second-order rules of validity of our own society. The positive law of a state is in fact the system of law which the courts recognize and apply in deciding the rights and duties of litigants. Now judicial recognition of rules is far from arbitrary, as we have seen. Initially, in our society, the courts look to the source from which the rules come, and they recognize four such sources. These are (i) customary rules, local or general; (ii) the enactments and ordinances of bodies such as local governments, government departments, etc.; (iii) judicial decisions of the higher courts, or rules derived from precedents; (iv) acts of the legislature. A rule from any of these sources is given initial recognition by the courts, but to receive operative recognition it must possess certain qualities or pass certain tests. In the case of (i) – customs, local or general – the tests are (*a*) that it must be of long standing and common knowledge, (*b*) that it must refer to matters of some importance, (*c*) that it must be consistent with the general principles of the legal system and the principle of equity. In the case of (ii) – local acts and ordinances – the criterion of *intra vires* must apply; i.e. they must be within the constitutional power of the body making the rules. But even after the application of these tests it is possible that rules from different sources may conflict. To meeet this situation, the courts recognize degrees of authority in the sources, and they accept that rules from source (iv) – legislation – prevail over all others.

This account of the second-order legal rules of our constitution is probably inaccurate in detail and is certainly too sketchy to serve as 'Instant Constitutional Law', but it may be sufficient to indicate the sort of function performed by second-order rules of validity. If a rule satisfies these second-order rules of legal validity (or similar ones) then it is legally valid or legitimate whether or not it is also valid morally.

So far, then, it has been argued that there is a distinction to be drawn between legal and moral legitimacy, and that a rule is legally legitimate if, but only if, it satisfies the relevant second-order criteria, and that the question of whether it is also morally acceptable does not directly arise at this point. But what is it for a rule of law to be morally legitimate?

3. 'LEGITIMACY' AS A MORAL CONCEPT

In answering this question we must draw an important distinction. The issue of moral legitimacy can arise over the existence of an entire legal system or it can arise over the existence of specific rules of law within the system. Thus we might say that a given system of government was morally illegitimate in that it imposed itself on the bulk of the people against their wishes, or we might say that a given rule of law was morally wrong within a system which was in general to be morally approved. In this chapter we shall be concerned more with the first of the questions, reserving the second as one of the issues to be discussed in the next. What, then, makes a system of government morally, as distinct from legally, legitimate or authoritative?

There seem to be two main views of what makes a government morally legitimate. In terms of the first, a government is thought to be related to its people as a guardian to a ward or a parent to a child. The government, that is, is regarded as exercising an authority, either inherent or bestowed by some divine power, and this authority is thought to be legitimate in so far as it is directed to the maintenance of a certain mode of life which includes the true good of the governed. This type of theory can be found expounded in Plato's *Republic*, and real-life examples of it can be found in feudal types of society or patriarchal societies. For example, the history of the Jews under Moses and Samuel suggests that political society in that period was based on or reflected such a theory. As always when dealing with real-life examples, modifications may be needed here. For while it is true that Moses, Samuel, etc. derived their authority from divine right it is also true that the relationship between the Jews and Jehovah was contractual, i.e. it was based on covenant. Similarly, in some medieval political theory it might seem that there is a divine right assumption about the nature of government, but the medieval theory of natural law tends to modify this. Despite all the modifications which must inevitably be made when we deal with actual constitutions, however, it is clear that there is an identifiable strand in moral thinking in terms of which governments are morally legitimate if in some sense they can be seen as the fathers of their people, whether these are the highly trained 'Guardians' of Plato's imagination or the 'Benevolent Despots' of the eighteenth century. Such a view would not make governments morally legitimate in the eyes of a liberal-democrat, however, for it is incompatible with granting the individual the right to influence policies by his own responsible decisions.

The second main view of what makes a government morally legitimate uses the analogy of the legal contract; government and governed are thought to be related as the parties in a contract voluntarily entered upon. This type of theory has antecedents in Greek thinking – Glaucon and Adeimantus suggest a form of the theory in Plato's *Republic* – but it is in the seventeenth and eighteenth centuries that the theory is to be found in a developed form. There are various historical reasons why during this period contract theory should become the preferred account of what makes a government morally legitimate. One reason is connected with the effects of the Reformation. The Reformation is sometimes described (by Protestant historians) as the rejection of authority in favour of individual conscience, and at other times it is described (by Catholic historians) as the breakdown of authority. The truth is rather that it was a time of the *division* of authority. Conflicting views existed about the proper interpretation of the Scriptures and about the proper principles for man's life in society. These conflicts led men to seek by reason a common ground on which all rational beings could agree, and we find an upsurge of speculation on the idea of a 'natural law' or 'law of reason' binding on men as men. It was in this context that the idea of a legal, rational contract between government and governed appealed to the imagination as the preferred account of what makes government morally legitimate. A second reason was the prevalence of war during this period – wars of religion and nationalistic wars. This again encouraged the search for rational principles appealing to men as rational creatures and useful for providing a basis for peace in the international field. Once more the analogy of the legal rational contract seemed appealing.

There was considerable difference among thinkers as to the interpretation of the alleged social contract. For example, the thinkers of the period do not agree as to whether the contract is between people who decide to form a society (a social contract strictly understood) or between the members of an existing society and a government, or both, but an increasing number of thinkers respond favourably at least to the general idea of a contract. They all take the view that a government does not hold office by virtue of some divine right but holds it because, and only in so far as, it performs a certain function in society. In other words, the various forms of contract theory stressed that the authority exercised by the government somehow stemmed from the will of the people, and that the people were entitled to withdraw their support from the government if it failed to fulfil the function for which it existed, or if it abused the authority by which it performed its function. It was to express this attitude towards government that the concept of contract was used, and it was in terms of the fulfilment of this contract that people saw their government as morally legitimate.

Taken literally, contract theory is easy to criticize, and there was an increasing volume of criticism of the theory from the mid-eighteenth century onwards. Historically speaking, it was said, the members of a society have never made contracts either with each other or with a ruler, and even supposing they had we would not now regard ourselves as bound by them. But even if the contract theory is not taken as expressing a view as to the historical origins of society but rather as expressing metaphorically the relation of ruler to ruled, we might still object to it; the metaphor suggests a relationship altogether too mechanical and legalistic. Nevertheless, there is a real point which is being expressed by means of the metaphor of contract: that the government may command with authority, and not just with power, so long as the members of the community consent to the general ends pursued by the government and the means by which they pursue these ends. It is in consent and not in contract that we must look for the theory of what makes government morally legitimate in our society. Whereas the term 'contract' is apt to suggest something completed once and for all, the term 'consent' can more easily be analysed as a continuous process, as an organic relationship of functional dependence.

The analysis of the concept of consent is also problematic, however, and many modern political theorists consider it unrealistic as an account of the relationship between government and subject. Certainly the concept must be used with precision. We may suggest, however, that a government rules by consent if and only if it is a necessary condition of its having a *right* to govern that those governed have expressed a wish or have otherwise indicated acceptance that it should govern.* It does not govern by consent in this sense simply because it in fact carries out the wishes of its subjects, or even where the fulfilment of their wishes is a necessary condition of its ability to rule effectively. Consent is essentially the granting to someone the permission to do something he would not have a right to do without such permission; and government by consent therefore implies that the government's *right* to govern (not its power to do so, nor its actual policies) is created by the expression of the wishes of the subjects.

To analyse the concept of consent in this way is to give it a strict meaning which will rule out many governments loosely called governments by consent. For example, benevolent despots generally rule by consent in the sense that the governed tolerate, or positively approve of, the policies of the despot. This was true of Elizabeth of England. But consent in this sense does not give us an account of what makes government morally legitimate. To have such an account, one which roughly fits the moral attitudes and views of our society, it is important to insist on the strict sense of consent.

* cf. J. P. Plamenatz, *Consent, Freedom and Political Obligation* (London: O.U.P., 1938).

But what does it mean to say that the government's *right* to govern is based on the people's having accepted the fact that it should govern? Or, to put the question in a more specific way, what is it that the people consent *to*? The answer to this question enables us to tie together our accounts of legal legitimacy and moral legitimacy. A government governs by consent, or is morally legitimate, when its people consent to the constitution, or (in other words) to the secondary rules operative in their society. A government, then, has authority, or is legitimately in authority, if but only if it is rightly constituted in terms of the constitution or the secondary rules, and if but only if the bulk of the people consent to the secondary rules. If these conditions hold, it is both legally and morally legitimate and can command with authority. This is the essence of democracy.

4. REPRESENTATION

The analysis of authority is not yet at an end, however, for the Western form of government is not simply democracy but (Swiss Cantons apart) *representative* democracy. What is it for one person to represent another, or to represent a multitude? The difficulty in answering this question lies in the fact that there are several different senses of the term 'represent'.* The first sense is that of *symbolic representation.* For example, a monarch or president may on certain occasions symbolize the majesty of the State. Again, an ambassador often has the function of symbolic representation in another country quite apart from other functions he may have as a representative. Thus, if an ambassador is withdrawn this not only effectively interferes with relations between two countries but also symbolizes coolness between them. A second sense can be called *ascriptive representation.* In this sense a person may have a legal representative who acts on his behalf, and a Member of Parliament represents

* The analysis of 'representation' is based on A. Phillips Griffiths, 'How Can One Person Represent Another?', *Aristotelian Society*, Supp. Vol. XXXIV (1960).

his constituents in this sense when he negotiates on their behalf with a government department. A third sense is *representation of interests*. A Member of Parliament may concern himself with the interests of a particular group in society such as trade unionists, and in general an official representative will concern himself with the interests of the people of whom he is the ascriptive representative. But of course he can represent the interests of those of whom he is not the ascriptive representative. A fourth sense can be called *descriptive representation*. In this sense any typical member of a group may be said to be a descriptive representative of it. For example, a Member of Parliament who comes from a working-class background may be said to be a descriptive representative of the working class. The point of having a descriptive representative of a class or group is that he will share its outlook and interests and have a more sympathetic understanding of its basic attitudes, and hence be better able to present its case and represent its interests. Fifthly, we might say that there can be *representation of values*. Thus a person can represent the values of a social group in that he embodies in his personal qualities and attitudes the characteristic values and aspirations of the group.

These are five of the most common senses of 'represent', 'representation', and 'representative'. All of them are relevant in discussions of representative democracy, and indeed there is often confusion in such discussions because people do not make it clear which sense of the term 'represent' they are mainly concerned with, perhaps because they do not realize that there are several senses. For example, one person might say that the working classes are no longer adequately represented in Parliament and another might reply that they have the same degree of representation as any other group and a third might say that a certain political party represented the interests of the working class. Such a discussion would tend to be foggy because the disputants were using (in order) senses four, two, and three. All these senses are important and relevant in different contexts, but it is essential to be clear which one is in question in a given discussion.

Again, the various senses of 'represent' are often confused in the discussions by students and college authorities of whether students should or should not be represented on governing bodies such as senates. Students often argue for representation on the grounds that their representatives should always be present to negotiate on their behalf (ascriptive representation) and that this will be more sympathetically done by one of their own number (descriptive representation). The university authorities, on the other hand, often fear that if student representation comes about the interests of students or (cynically) their values will be over-stressed in senate deliberations. It may be, however, that the real case for student representation on the main governing bodies involves the first sense – symbolic representation. In other words, it may be that students feel that they should be recognized as full and responsible members of their institutions and that this aspiration would be symbolized by having one or two of their number sitting with the teaching staff on governing bodies. The fact that their practical contributions to discussions would necessarily be restricted by their lack of experience (although this would not be true of all matters) would be irrelevant to symbolic representation.

Let us now try to link the senses to provide the first step in an account of how democracy can be representative. The basic sense is the second – ascriptive representation. One person is the legally constituted agent for a group of people. He acts in a role which is created by the secondary rules of society, and as incumbent of the role he is in authority to the extent of the rights and duties which define the role. For example, a Member of Parliament is the legally constituted or ascriptive representative of a constituency and can act on behalf of his constituents. This is his role and the rights and duties which define it are laid down by the secondary rules of our society. By occupying the role he is empowered or enabled to act in various ways which would otherwise be improper for him. But the main aim of his actions in this role is given by the third sense of 'represent'. In other words, the Member of Parliament acts in his role to further the interests of the people of whom he is the ascriptive representative. (There is more to his role than this, as we shall shortly see.) This third sense is essentially involved, in that an ascriptive representative who consistently acted against the interests of the people he was supposed to be representing might be said not simply to be a bad representative but not really one at all. Now an ascriptive representative who was also representative in the fourth sense (i.e. who was also a descriptive representative) would be more likely to be vigorous and effective in representing interests. This is a contingency with limited relevance since, at least in parliamentary affairs, a representative should be concerned with more than the interests of a sectional group of his constituents, and it is conceptually impossible for one person to be the descriptive representative of more than a few groups. We also expect that our ascriptive representative will represent our values. This point goes without saying in the internal affairs of a country where there is probably some uniformity in the values which are largely acceptable to the bulk of the people, but it becomes more important, as we shall see,* in the foreign affairs of a country. And perhaps on certain occasions the representative may, in the first sense, symbolize something about the people he represents.

Now what we have just given is only the first step in the account of representation. There are still several factors to be taken into account. In the first place, a Member of Parliament does not simply represent the interests of his constituents but also concerns himself with the affairs of the country as a whole (and, of course, the same will be true *mutatis mutandis* of other forms of government as well as parliamentary democracy). This was what was meant when we said that there is more to the role of a Member of Parliament than simply furthering the interests of his constituents. Parliament is also a forum in which the main issues of the day are debated and discussed, and each representative has the additional role of spokesman for the community as a whole in terms of

* Chapter 5, section 3.

certain broad party policies. We can say, then, that to account for certain phenomena of representative government, we must bear in mind that each ascriptive representative has the additional role of party member, and that that role is in the end justified, it is believed, in terms of the interests of the country as a whole which are held to be furthered by the debates and conflicts of party politics.

In the second place, the main decisions of internal and foreign affairs are taken not by all the ascriptive representatives but by a group which, by means of party machinery, ensures that its decisions are given majority parliamentary support. This group, the Cabinet or 'Government' in a narrow sense, acts *as a group*. We also speak of 'the State' acting on behalf of the people. How can a *group* be the representative of another group?

The answer to this question requires the introduction of the concept of a corporate person. A corporate person may be defined as a group of persons, individual or corporate or both, so organized that they are capable of exercising rights and performing duties pertaining to the ends of the group as a whole and not just to those of its individual members. When the government is acting as a whole it is acting as a corporate person and its rights and duties pertain to it as a corporate person. The form of speech which attributes actions to the government as a whole, as opposed to the specific ascriptive representatives in it, is therefore not metaphorical, since it can be given conceptual justification in the notion of a corporate person.

5. CORPORATE PERSONALITY AND RESPONSIBILITY

Although the attribution of corporate personality to the government may explain the sense in which it can be said to act as a single unit, and thus be able to represent another group, it nevertheless brings a new problem into focus: that of how *responsibility* can be attributed to the government as a whole. Indeed, this is only one instance of the general problem of the sense in which responsibility can be attributed to any group or collective. If the argument of Chapter 3 is sound, then responsibility belongs to the adult individual. How then can we attribute responsibility to a group or corporate person such as a government?*

One answer suggests itself straight away. A group can be regarded as responsible provided that it is possible to reduce the decisions of the group to the decisions of consenting members. In other words, provided that it is possible to see the group decision as resulting from the sum of the individual decisions, we are justified (it might be said) in attributing responsibility to the group decision.

A difficulty with this solution is that it does not seem to be the case that all group responsibility can be reduced in this way to the responsibility of the individual members. For example, if a club owns premises it may be responsible for what happens on the premises. But it surely does not follow that every individual member of the club is responsible for what happens on the premises; it is the club as a corporate person which is responsible, and in this type of case the responsibility of the group does not seem to be reducible. Again, if a limited company goes bankrupt it does not follow that every director of the company is held responsible. Finally, if 'the government' acts it does not follow that every member of the government is responsible for what the corporate person has done. It seems, then, that it is not true in all cases that the responsibility of a group is reducible to that of its individual members.

The answer to this difficulty is to distinguish between moral and other responsibility, such as legal or political. All we are obliged to hold in order to maintain the doctrine of individual responsibility, as it was expressed in Chapter 3, is that the *moral* responsibility of a group must be reducible to that of its individual members: we can allow that the *legal* responsibility of some groups may be irreducible. And the apparent counter-examples to the initial thesis seem to involve only legal or political responsibility.

* I have discussed the nature of corporate action and responsibility in *Government Action and Morality* (London: Macmillan, 1964).

It must be noted, however, that the moral responsibility of a corporate person or group is not *directly* reducible to that of its individual members; they are morally responsible for what the corporate person does in virtue of their acceptance of, and action in, the roles which constitute the corporate person. In other words, their decisions as individuals do enter the analysis, but in a relatively complex way. In fact, we might say that individual moral responsibility is involved in corporate action in three ways. Firstly, the rights and duties or roles which constitute the corporate person have been created, developed, or maintained by the decisions of individuals. If a corporate person tends to produce actions with a characteristic moral quality this will be partly because the decisions of the individuals who created it are to be morally praised or blamed. Compare, for example, the moral decisions of those who created 'The Gestapo' with those who created 'Oxfam'. The fact that the individuals who created the corporate person may not now be acting in its roles does not absolve them from moral responsibility for creating it. In the second place, some individual person freely decided to become a member of the corporate person, in the sense that he accepted a certain role or set of rights and duties. The morality of 'role-acceptance' is, of course, complex – all sorts of pressures may force a person to join a certain group – but we can still maintain that the decision to join it is basically an individual decision and carries moral responsibility with it. And a person can resign if he disagrees violently with the action he must take as a member of a group. Finally, a person can bring various moral qualities of his own to his actions as a member of a corporate person. For example, the corporate person 'The Home Office' has sometimes been morally criticized for the behaviour of its Immigration Officials. This is because some of them are alleged to bring to their actions as members of the corporate person undesirable moral qualities, such as rudeness.*

In sum, then, we may say that the concept of a corporate

* Some of the issues of this paragraph are discussed in more detail in Chapter 6, sections 2 and 4.

person or constituted group has been introduced to account for the fact that a group or collective, such as a government, can act on behalf of other groups or individuals. The concept of corporate personality gives rise, however, to the problem of corporate or group or collective responsibility. My argument has been that, whereas the legal responsibility of a corporate person may not be reducible to that of individual persons, the moral responsibility is so reducible, but in an oblique way. The concept of corporate action and responsibility, which is necessary for political and legal analysis, is therefore compatible with our ordinary ideas of individual moral responsibility.

6. CONCLUSION

The argument has been that a person is in authority if he acts in a role which has been created by a set of what have been called secondary rules. To say this is to say that the role has legal authority. But it also has moral authority in so far as the people governed by the authority consent in some sense to the secondary rules. A person becomes a representative authority in so far as the secondary rules empower him to speak on behalf of a group (constituents or the like) and to further their interests in his representation, displaying values in his conduct which are shared by his group. A group of such representatives can form themselves into a corporate person which, like the individual representative, can also act on behalf of others, furthering their interests and displaying values which the majority can share. But, whereas the legal responsibility of a corporate person may not be reducible to that of its individual members, its moral responsibility must be so reducible if we are to sustain the idea of Chapter 3 – that the individual person is the ultimate bearer of moral responsibility. And it is reducible, but in a complex way. We now have the general outline of representative democracy and why it is authoritative for people in our type of society. In the next chapter I shall add further important points to the account.

5

The Morality of Government Action

The position so far reached may be described as follows. Human beings, because of the nature with which they have been endowed and the environment in which they must live, will tend to accept certain rules for their guidance. These rules can all be subsumed under the wider principles of utility, equality, liberty, and fraternity (if we may for convenience regard the last as a principle) which in their turn may be said to presuppose the basic principle that what ultimately matters is the individual – that persons ought to be respected as ends. The social phenomenon generally indicated by the term 'authority' required further principles and concepts for its adequate analysis. In particular, the idea of secondary rules was needed to explain the concept of legal authority, but we also had to say that authority was morally binding on people only in so far as they accepted the secondary rules. If we add to this the idea that certain people in roles rightfully constituted by the secondary rules, and certain corporate persons, again rightfully constituted by the secondary rules, represent the people who consent to the secondary rules, we have the main outlines of representative democracy.

1. THE PUBLIC GOOD

We can now add further details to this picture if we ask the question: what is the relationship between law and morality? This question is characteristically raised in a way which, while it is quite intelligible within the limits of a particular kind of controversy, is from a wider point of view misleading.

The question is often posed in the form: should the law enforce morality? It is misleading to pose the question in this way since, if the analysis in this essay is on the right lines, the law should not enforce anything else. The general lines of our argument have been towards the conclusion that the term 'morality' can be applied to the principles of social organization of a community, and 'the law' is a particular set of these rules and principles which become 'legal' in so far as they satisfy the rules for legal validity. Hence, the controversy often identified as that between 'law and morality' is strictly a controversy not about *whether* the law should enforce morality but rather about *how much* morality the law should enforce. It is the controversy as to whether there are areas of a person's life in society on which the law should not have a view and, if so, how this area may properly be identified. And in this form we shall later discuss the problem. In the meantime, however, it will be profitable to examine the area in which morality and law overlap. They overlap over the rules which are basic to sustaining the life of a society. Let us consider what form these basic moral rules or principles take when they are legally incorporated and become the subject-matter of political debate.

People tend to accept a set of rules, we said, which are roughly characterized as rules of utility, that is, as leading to the greatest possible good – happiness, or the like – of the majority of the people in the society. When this is put in the customary language of traditional political theory it emerges as the formula that an important function of government is to further the 'public good' or the 'national interest'. Such terms are often criticized as being too general and vague to be helpful, but if carefully used they can be useful shorthand expressions.

There are three main elements in the idea of public good. In the first place, the public good has often been thought of negatively as a matter of protecting the interests of the inhabitants of a given political unit or territorial area. This means that the citizens must be defended against invasion or attack from outside and that the Government may therefore maintain armed forces and levy such funds as are needed for their support. This has always been regarded as a function of the Government, although opinions may differ widely on the precise meaning of 'adequate defence'. It is equally obvious that public good requires the protection of the citizens from attack from within the State itself. This may mean protection from subversive elements within the State which would seek to undermine it, or it may mean protection from criminal elements which would attack the life or property of the inhabitants. At all events the conception of public good clearly involves the maintenance by the State of some body with the function of preserving peace within as well as without the State. The protective function of the State can clearly be extended much further. For instance, the citizens require protection against disease, so there is room for provision by the Government of systems of sanitation and for the development of various public health regulations. Again, in any society in which there are mechanized forms of transport, traffic regulations are necessary to safeguard the lives of the citizens. It will be generally agreed that in these cases, and there are many similar ones, the Government may impose such rules and regulations on, or require such services of, its citizens as are necessary to further the end of public good, understood negatively as a matter of protecting the interests of the people.

The function of the Government is obviously not exhausted by this negative conception: more is involved in the public good than protection, no matter how extensively the protection extends. Indeed, there is already implicit in the conception of protection a second more positive function of government. If we consider the protection of a country against disease, it is clear that no line can be drawn between this and the positive furthering of the health of the inhabitants. A lasting freedom from disease can be had only if the inhabitants of a community are basically healthy, and the health of a large number can be secured only by the provision of welfare services and so on. The concept of public good is therefore to be analysed negatively into protection against hostile elements of any kind and positively into the provision of various services which will lead to the well-being and prosperity of the community as a whole.

This second function of the State is one which has come into prominence in this century, although it must be present in some form wherever there is a political society. But recently many functions formerly fulfilled by private bodies have been taken over by the Government, and the pressure on the Government to extend its social services is increasing. Not only do we look to the State to provide us with health services, an educational system, roads, postal services, and so on, but we also expect it to encourage industry, to ensure a basic standard of living for everyone, and even to take over most of the functions of the family by the provision of various services for children and the aged. Cultural life also is encouraged by the Government or government-sponsored bodies, and many people would like to see the Government playing an even greater part in developing the cultural life of the nation. Opinions differ, of course, on how far the Government is justified in going in the direction of positively furthering the interests of the citizens. Some people feel that it has already gone too far in extorting money by taxation to provide certain services, or they argue that in certain areas government interference will defeat its own ends by stifling the life it is seeking to encourage. Others again hope to see the range of government activity greatly extended till it permeates every area of human life.

The two elements into which 'public good' has been analysed are sometimes expressed by means of the concept of freedom. The negative sense of public good – protection – can be expressed in terms of 'freedom from'. Thus protection against outside invasion can be called 'freedom from oppression' and against civil disorder 'freedom from fear' and against disease 'freedom from disease'. The more positive sense of public good – social welfare of all kinds – can be expressed in terms of 'freedom to'. Thus, the provision of educational services gives the freedom to engage in intellectual pursuits. It is doubtful, however, whether any clarification of the concept 'public good' can be achieved by even more ambiguous than the former. The term 'freedom' analysing it in terms of 'freedom' since the latter term is in a political context does little more than signal the presence of a favourable emotion and it is therefore of no value in the analysis of any other concept. It is preferable, then, to break the concept of public good into the concepts of public protection and public welfare, both of which terms admit of a more concrete analysis.

There is, however, a third element involved in the public good and the third element is at least as important as the other two. The public good requires that there should be an independent arbiter between conflicting group interests; and this is the third of the functions performed by the Government. In any society, and especially in a complex one such as our own, there are bound to be groups with competing or even conflicting interests. The interests of consumers, for instance, may conflict with the interests of retailers: the former want to pay as little as they can for a given commodity while the latter want to make the highest possible profit on the same commodity. Again, the interests of an industrial firm which wants to dispose of waste products cheaply by pouring them into a river may conflict with the interests of those concerned with preserving the beauty of the countryside or with fishing the river. Employer and employee, landlord and tenant, television advertisers and viewers, farmers and house builders, may all conflict from time to time, or all the time, over specific details or general policies. In such situations the Government acts as an independent arbiter deciding issues in terms of the public good.

It may be asked, however, by what criterion the public good is decided. After all, whichever way the Government decides in a dispute will produce benefit for one part of the public, and it is to be doubted if there is one good in which every member of the public can share. It may be suggested that the good in terms of which the Government will decide an issue is the good of the majority. But this is not always the case: sometimes a minority may be favoured and sometimes all may be required to make sacrifices. The 'public' whose good is furthered cannot, then, be reduced to the good of any group, majority or minority, within the community at a given time. It cannot even be the individual goods of the people who, taken collectively at any given time, make up the community. This suggests that the 'public' whose good is furthered by the Government is the whole continuing social community. Just as an adequate conception of government action requires the use of concepts which cannot be reduced to concepts about the actions of individual persons, so an adequate analysis of public interest requires the use of concepts which cannot be reduced to concepts about the interests of individual citizens or groups. To say this, of course, is not to say that there is a mysterious organism, 'the public', over and above the individuals who make up the public, or to assert the existence of a mysterious entity over and above the individuals who make up the Government.

To argue that the good of the public is not just that of the individuals of whom at any given time the public consists is not to suggest that the Government may impose some conception of intrinsic good or natural law with no relation to what individuals within the society actually want. In some political traditions it may seem permissible for a government to pursue a single abstract ideal for human life in society, but within our own tradition this forms no part of the Government's function. The point is rather that the Government must develop the implications of what the public actually want in order to elicit an enlightened conception of what in the long run they really want. Independent arbitration in the interests of the public good may therefore involve decisions designed to preserve institutions or to further policies which produce no *immediate* benefit for anyone. Assuming the acceptability in principle of this account of public interest or good, the object of independent arbitration on the part of the Government, let us go on to examine in more detail the nature of the independent arbitration.

In discussing the nature of government it is customary to distinguish three organs – Executive, Legislative, and Judicial – which are jointly responsible for the processes of government, and an adequate account of the Government's function as an independent arbiter must refer to all three. The Legislature is obviously concerned with introducing new laws and regulations, and these will act as checks and balances to existing sectional interests or they will adjust the social order to accommodate the arrival of new institutions. In general, then, the function of the Legislature as an independent arbiter is that of holding in continual tension the constantly changing forces of society, and this is done by the enactment of new legislation or the readjustment of existing legislation.

The social situation is complicated, however, by the fact that one of the main forces at work in it is the Government itself, understood narrowly in this context as the Executive. For example, in pursuing the public good, interpreted in either its negative sense of public protection or its positive sense of public welfare, the Government as the Executive may clash with a sectional interest. This may be either the interest of a powerful section, such as a large industrial concern, or that of a single individual. For example, the Executive may wish to build a road through someone's house, and the person in question may feel that he is being unfairly treated because the road need not go through his house, or because he is not receiving adequate compensation. In such a situation the Executive does not necessarily prevail. The aims of the Executive are not automatically identified with the conception of public good but are rather regarded as a further sectional interest the claims of which are to be balanced against those of the other person involved. To decide the issue the Government interpreted as the Judiciary may be called in, and the function of the Government as arbiter is that of deciding which of two sectional interests may prevail; both the private person in question and the Executive are legal persons. Independent arbitration is still a function of government, then, even when one of the parties is an organ of government.

The Judiciary decides the issue mainly in the light of existing legislation. But not all of the relevant law may have issued from the Legislature or, if it has, it may have issued from a Legislature concerned with a very different social framework. In such a case the Executive can reassert itself as an organ of government and insist on having its own way. This it can do by introducing new legislation which will enable it to win its case. The Executive, of course, cannot by itself pass the new legislation, but it can be introduced by the Executive, and, by the system of Party Whips, it can be forced through the Legislature. Naturally, such a procedure seems at first sight to be at variance with the conception of the Government as an impartial arbiter between conflicting sectional interests, but in some cases – and the expedient is only infrequently used – it may be justified by the conception of public good. After all, it is not always morally right to hold slavishly to a moral rule when the consequences of doing so will be disastrous.

Special legislation is sometimes introduced in cases where the Executive is not one of the contending parties. For example, in 1900 a majority of members of the Free Church of Scotland decided to join with the Church of Scotland, despite the significant doctrinal differences. The minority group of the Free Church claimed the funds on the grounds that the majority had departed from the doctrines which defined the nature of the Free Church. The House of Lords upheld this view, adopting a strict view of the nature of corporate personality. On this occasion the Government decided that the public good would not be furthered by this decision and, acting as a relatively detached arbiter between two parties of which it was not one, it passed a special Act of Parliament which divided the funds equally between the two contending parties. The Government, then, furthers public good not only by protecting its citizens and improving their welfare but also by arbitrating between sectional interests. 'Government' in this sense, however, is a function of the three organs, the Legislature, the Executive, and the Judiciary.

We have been giving an account of how government, as a body constituted by secondary rules, is competent to give legal backing and dynamic implementation to the basic primary rules of society. A government can implement these basic primary rules with authority in so far as it is rightfully constituted by the secondary rules, and the majority of the people consent to the secondary rules. The idea of consent, it has already been argued, enables us to understand the moral authority of government.

But the concept of consent is useful in rather a different way to account for the moral authority of a *particular* government. For the aim of government is to give effect to the principles of utility, equality, liberty, and fraternity on a scale beyond that of the individual or small group of individuals; it is to compensate on a large scale for our individual deficiencies. Now particular governments will have rather different ideas on how this may best be done – that is one of the reasons for the existence of party politics – and most people have views on how far they think a particular government or set of policies has been or is likely to be effective in realizing what they think they want. Ideally everyone would participate in the governmental processes and the idea of consent in this second sense expresses this.

It is obvious that the theory of consent even in the second sense is unrealistic if it is regarded as an attempt to mirror the facts of government in a large modern state. Every individual cannot participate directly in the processes of government, and, even were it practicable to hold a referendum on every major issue, it would not be desirable, since many issues are of too technical a nature for the average member of the community to hold informed opinions on them. Indeed, most people remain comparatively uninterested in the vast majority of political issues. Nevertheless, consent theory does express a truth which is fundamental to the liberal-democratic tradition: the people are not puppets to be manipulated at the whim of the Government. Sooner or later the Government of a democracy must go to the people, who then have the opportunity to participate indirectly in the processes of government by voting into power that party the general aims of which they regard as most nearly fulfilling their own. Further, even when the Government is in power, the members of the public can still participate indirectly in the processes of government to the extent that they can voice their criticism of particular government actions, or draw the attention of the Government to particular abuses which they think ought to be remedied. This can be done by letters to Members of Parliament or newspapers, or through local party organizations. To say, then, that in the last resort the Government rests on the consent of the people it governs may not be to assert that there is a direct connection between the will of the people and the legislation which a government actually introduces. It is, however, to assert something important: that the range of choices open to policy-makers is restricted by the necessity of obtaining a periodic mandate from the people. This is the truth which is expressed in the claims that those governed can participate in the processes of government and that to the extent that this is possible the existence of particular governments may be regarded as legitimate.

2. LAW AND MORALITY

It was maintained at the beginning of the chapter that the question 'Should the law enforce morality?' was misleading, since the law should do nothing else. A more accurate way of putting the question is, 'How much of morality should the law enforce?' or 'Is there an area of morality on which the law ought to be silent?' Let us turn now to this topic. In fact, there are really two issues to be discussed. The first is whether the law should enforce any positive ideals of the good life which extend beyond the basic rules and aims expressed in the idea of the 'public good'. The second is whether there are any moral vices which the law should discourage even although they are not necessarily contrary to the public good. Let us consider the first issue.

At first sight it seems as if it is no part of the Government's function in our tradition to pursue ideals. We described the Government's function as the protection of its citizens against some of the uncertainties of human life, the creation of welfare services, and the provision of arbitration among competing sectional interests. If the Government's function is so described, two arguments, a more and a less radical, may be urged against the view that a government can pursue ideals.

The more radical argument is to the effect that the pursuit of an ideal is actually incompatible with the function of government as I have described it. The Government, it may be said, has been depicted as properly concerned only with such matters as arbitration, which affect the community as a whole. To pursue an ideal, on the other hand, whether it be Christian or Communist or Fascist, will involve the elevation of some sectional interests at the expense of others, to the consequent detriment of the arbitral function of government.

This objection is valid, however, only on the assumption that if a government interests itself in ideals it will be concerned exclusively with one ideal. Thus, if the Government pursued exclusively the ideals adopted by, say, music lovers, sports fans, or religious enthusiasts then it might legitimately be argued that it was failing in its function of independent arbiter. It is possible, however, for a government to pursue, or encourage its citizens to pursue, a *variety* of ideals without in any way impairing its function as an independent arbiter. The radical objection fails, then, if it is stressed that the pursuit of ideals by a government does not commit it to the ideals of any one sectional group.

But, although the pursuit of ideals in this manner may not actually be incompatible with the function of government, some theorists might argue that a government is not really a suitable body for the encouragement of ideals since it is cumbrous, bureaucratic, and official, whereas the pursuit of ideals is essentially a matter for private initiative and individual self-realization.

This objection misconceives the manner in which a government can pursue ideals. The suggestion is not that there should be a government department concerned with ideals, but rather that a government should encourage in many different ways the pursuit of ideals by its citizens; it should create the political atmosphere in which a plurality of ideals of the good life can be followed by the citizens. For example, it should encourage the correct use of leisure by the provision of educational facilities and it should encourage the arts by the commissioning of public works.

The main argument for suggesting that a government should encourage the pursuit of ideals so understood is simply that such pursuits are good in themselves; they satisfy the non-material needs and aspirations of the citizens and it seems a legitimate governmental function to do this. A second argument for the view that a government should promote ideals of the good life can be formulated by developing the point already stressed in answering the objections – that a government should encourage a *variety* of ideals. A. N. Whitehead* has maintained that progress is founded on the experience of discordant feelings. He points out that the Chinese and the Greeks both achieved perfections of civilization, but that 'even perfection will not bear the tedium of indefinite repetition. To sustain a civilization with the intensity of its first ardour requires more than learning. Adventure is essential, namely, the search for new perfections.' Now it is in the idea of a 'search for new perfections' that we can find a positive argument for the view that a government should encourage the pursuit of ideals, for such encouragement fosters the creative impulses which sustain a society. Lacking the self-criticism and the imaginative vision which come from ideals a society stagnates; its science declines, its literature becomes shallow and its social services lack humanity. Such a condition of society is clearly undesirable: it is therefore part of a government's function as promoter of the *public good* to prevent its development by encouraging the pursuit of ideals.

Let us turn now to the second of the two issues involved in the question of how far the Government should enforce morality, the question of whether the law should be used to stamp out 'vice' or 'sin' as such, that is, moral wrong-

* A. N. Whitehead, *Adventures of Ideas*, p. 332 (Cambridge, 1933).

doing not obviously detrimental to public good. This issue has gained public prominence in the last decade or so as a result of several legal cases – such as that of the *Ladies' Directory* – and the Wolfenden Report on homosexuality.* The controversies tended to be confused, partly because they were sometimes discussed in the form of the question whether 'sin' should also be counted as 'crime', and irrelevant religious and anti-religious factors complicated the debate. Let us ignore the religious factors, however, and consider the other arguments for and against the view that the law ought to be directed against vice as such.

There is one very general argument against the view that the law should concern itself with vice as such, and it concerns the consequences of legislation against vice. One consequence is that the risk of blackmail is greatly increased; but the evils of blackmail, both in the personal distress it causes and the security risks it can create, are notorious. A second consequence concerns the methods of enforcing legislation directed at the vice of consenting adults in private. Whatever the moral rights and wrongs of certain forms of conduct we ought to be alive to the dangers of allowing the law to intrude too far into our private lives. The methods of obtaining evidence, and the victimization by the police which has sometimes been alleged to occur, seem more unsavoury than the alleged vice against which they are meant to be directed. Mill sums up this argument when he writes: 'We should be glad to see just conduct enforced and injustice repressed, even in the minutest details, if we were not, with reason, afraid, of trusting the magistrate with so unlimited an amount of power over individuals.'† Mill is here speaking of justice and injustice, but his argument will apply *a fortiori* to what is called 'vice'. The argument seems a strong one against the view that the law should concern itself with private vice not demonstrably harmful to public good.

* There is a discussion of this and connected issues in H. L. A. Hart, *Law, Liberty and Morality* (London: O.U.P., 1963).

† J. S. Mill, *Utilitarianism*, Chapter V (Fontana edition, p. 303).

Nevertheless, it may be that despite the unfortunate consequences of legislation against vice there are other factors which ought to have precedence in considering the function of law. One of these factors is the protection of people against the unwitting consequences of their own immoral conduct. For example, it might be said that people ought to be protected against the dangers of drug-taking, and that this can be done to some extent by legal threats against possessing or using certain drugs.

One difficulty attached to discussions of this issue is that drugs vary a great deal in their effects, such as how far they are addictive. But let us assume that for the moment we are talking about highly addictive drugs, such as heroin. (We shall later consider the different issue of non-addictive drugs.) One argument in favour of legal restraints on the use of addictive drugs is a simple utilitarian one. The addict is a dangerous person to have loose in society since he is liable to turn to other forms of crime to ensure the supply of his drug. Moreover, the treatment and rehabilitation of an addict is costly to the community. Hence, there is certainly an argument in favour of legislation against the use of addictive drugs, but it is a straightforward utilitarian one and does not require to be seen as legislation against vice as such.

Many people feel, however, that, quite apart from the fact that addicts become a drag on the community, we ought to protect the potential addict against his own rash judgements. If this is admitted we seem to be ascribing to the law a paternalistic role. Can we justify this role in terms of the general viewpoint of this book?

It is worth noting, to begin with, that the paternalistic function of the law is not restricted to the cases of so-called vice. For example, there is a whole range of legislation concerned with provisions for the protection and education of young people. Again, in a welfare state there is a great deal of legislation designed to compel people to secure themselves against sickness, unemployment, old age, and so on. Hence, there is nothing particularly surprising or novel about the paternalistic function of the law. And if the general moral viewpoint of this book is accepted – that equality, liberty, fraternity and, in general, respect for the individual, are the basic values of our society, and that a government ought to further these in its policies – then we have a justification for paternalistic legislation. It is justifiable if it will protect the individual not only against attacks on these values by others, but also to some extent against his own misguided or ill-informed disregard of them. In other words, to the question of how much, or what aspects, of morality the law should enforce, we can answer that it ought at least to enforce those aspects which prevent a person engaging in activities which bring about a deterioration of his personality which is not easily, if at all, reversible.

There are two difficulties in this position. Firstly, the English law against suicide was removed in 1961 with general approval, yet this law surely was designed to protect people against their own irreversible decisions. The answer to this objection is that people presumably know what they are doing when they commit suicide, whereas the trouble with drug-taking is that it may be started without full knowledge of its consequences. Secondly, it may be objected that, even if people should in some matters be protected against themselves, it is not clear that the sanctions of the criminal law are appropriate for the purpose. There is much to be said for this objection. The criminal law is perhaps more appropriately directed against the possession or selling of dangerous drugs than against addiction as such. But it is by restricting the availability of dangerous drugs that the criminal law protects people against themselves. Hence, the objection does not disturb the general contention that it is consistent with the present values of our society that the law should take a paternalistic interest in certain aspects of private life.

The argument so far, however, has not suggested that the law should concern itself with conduct, such as homosexuality or non-addictive drug-taking, which many people would consider as 'vice' but which cannot plausibly be said to produce deterioration of individual personality. In the debate over the Wolfenden Report, however, some people introduced an argument which did have such implications. They maintained that irrespective of the effect that altering the legislation would have on the interests of homosexuals and young persons, the law ought to forbid homosexual practices because they were alien to our way of life, as shown in the fact that most people view them with repugnance.

Now if this argument is claiming that homosexual practices, even between consenting adults, will somehow *destroy* our society then it is a form of utilitarian argument, and perhaps an implausible one. But the argument is sometimes a more moderate one, saying only that the legalizing of homosexual practices will *alter* our society, will change its values, and that the Government should prevent this from occurring. In a somewhat similar way it has been argued that radical alteration in divorce laws or legalizing the popular use of certain non-addictive drugs will change the values of our society; or that unrestricted immigration will do this and ought to be stopped even although the immigrants perform valuable functions in the division of labour. People who argue in this way are drawing on the idea that a government is a representative not simply of interests, but of values as well. They are arguing that the Government is the trustee not only of the national purse but also of an inheritance of values.

There is a good deal to be said for the argument in general, whether or not one agrees with particular applications of it. It is in accord with the view that in a certain sense a government can pursue ideals of the good life as distinct from simply furthering material interests. Again, it is an argument based on a realization of the importance of social identity. People must know what sort of behaviour to expect of others in their daily lives and they must be able to feel akin to others. In short, in a healthy society there must be a moral consensus. This point was touched on in our discussion of fraternity. Properly interpreted, then, this argument makes a plea for conservatism in social change (a topic later to be explored in more detail) but does not supply new grounds for making alleged 'vice' as such illegal. The argument is often brushed aside by revolutionaries and liberals alike, but perhaps there is something to be said for such Burkean considerations in an age of impatience.

In view of the fact that discussions under the heading of law and morality are often heated it may be helpful to re-state the general lines of the argument in a different way. The usual liberal approach to the question is to deny that the law should concern itself with morality as such and to maintain that it should concern itself only with preventing harmful consequences or promoting good consequences to the public in general. It is not infrequent, however, for the same liberal writers to proceed in another chapter to ascribe to morality the function of preventing harm and promoting benefit. To fulfil this function, then, is *ipso facto* to be concerned with morality as such. Moreover, it is generally admitted that the law does have a legitimate function in furthering other moral values such as equality and liberty. In view of these facts, there seems nothing except a certain squeamishness to preclude a liberal from admitting that the law not only may but ought to promote moral values *as such.* There would appear to be nothing objectionable in this to liberals when it is appreciated what promoting moral values means.

A question does remain, however: should the law concern itself with *all* aspects of morality? And this is really the question debated under the heading of 'law and morality'. My argument has been that as far as the creative side to morality is concerned – the sphere of ideals – the law may legitimately provide encouragement, although it ought to direct itself to variety rather than uniformity. As far as the preventative side is concerned, the desirability of legal interference is limited because the methods of enforcement are likely to create more harm than they prevent. But if the consequences of certain practices are very harmful, even if only to the people concerned, the law may nevertheless have the paternalistic function of stepping in to protect them against themselves. Such a view may sound shocking to the entrenched liberal, but it really raises no principle which is not involved in the provision of compulsory insurance, etc. As for practices which are not now regarded as harmful and which, as a result, are not considered as 'vice' by the majority of people, it may be that the law does slow down social change. But in so far as this gives people the opportunity to understand what is involved and have sympathy with it then the law may here have the function of in fact aiding long-term social assimilation. Thus, the protracted debate over homosexuality has had the result that the public in general now have much more understanding of and sympathy with the position of the homosexual than formerly. Perhaps a similar process is taking place over the use of non-addictive drugs.

3. MORALITY AND INTERNATIONAL AFFAIRS

So far in this chapter we have been concerned with the internal affairs of a country, and have tried to show how certain of the basic organizational principles become the laws or legally enforced rules of the society. In this sense it can be said that a government provides a means whereby a system of morality can be operative on a large scale. But a particular country is only one unit in a larger international sphere. A problem therefore exists about the moral relationships among states as well as within states. Can a given state have a moral concern in an international field?

The reason for saying that it ought to have a moral concern is that morality has no national frontiers because its field is the social relations of persons as such. As Hume puts it, '. . . we give the same approbation to the same moral qualities in China as in England. . . . They appear equally virtuous, and recommend themselves equally to the esteem of a judicious spectator.'* If, then, we say that the Government ought to have a moral concern with respect to the international sphere we are really saying that it is organizing its people so that they may fulfil their moral duties to the

* David Hume, *Treatise on Human Nature*, Book III, Part III, Section 1.

members of this international society. Just as a government acts as what may be called a 'moral intermediary' on the domestic scene, organizing people into the functionally significant relationships which give rise to their duties, so it can act on their behalf as a moral intermediary in the international field and enable them to fulfil their moral duties to humanity at large.*

There are many difficulties attached to such a view. The first is that if we hold that the Government ought to act as a moral intermediary in the international field, it will follow that it may at times be obliged to act in ways which are not obviously in the national interest. And to many people this is a repugnant idea. They would argue that the Government represents the national interest, as a trustee, and that it is failing in its function – failing in trust – if it does not promote the national interest on all occasions in the international field. Indeed, one of the main aims of government actions in domestic matters emerged earlier in the chapter as that of furthering the public interest. But the difficulty which now arises is that whereas *domestically* the conceptions of the Government as a representative of interests and as a representative of values are not obviously incompatible, since it is in the national interest that the Government should represent our values by respecting its citizens and their moral ideals, *internationally* the two conceptions often seem to pull apart, and tensions arise between a government's obligations to its own citizens and to those of the world at large. It follows, then, that since representation both of values and of interests is involved in the conception of a moral intermediary, there are serious difficulties to be faced by its extension to the international scene. These difficulties depend on the acceptance of the two premises: that following the dictates of morality, or representing values, in the international field is not always in the national interest, and that a government must pursue the national interest at all costs. Let us investigate the premises in order.

* I introduced the idea of government as a moral intermediary in *Government Action and Morality* (London: Macmillan, 1964).

It is probably true to say that in general if a government takes on itself the representation of moral values in the international field it will also to a great extent be representing the national interest. 'Honesty is the best policy' may be in its own way as true among nations as among individuals. It is possible, however, that situations may occur in which representation of the national interest does require the abandonment of moral standards in policies or their enactment. Hence, the Government cannot both represent moral values and act on every occasion in ways which national interest demands. A lot depends, of course, on how long 'long-term interests' are allowed to be; and no doubt it can be held as an act of faith that policies required by morality will in the longest of all runs turn out to have been in the national interest. It is doubtful, however, whether it is legitimate to permit the term 'national interest' to be indefinitely extended: to be useful in political theory the term must have a defined range of application. Moreover, it is not acceptable to argue that a government may act as moral values demand so long as national interest is not impaired. It is a necessary truth that a government cannot recognize the claims of morality for some of its international actions and deny them for others. Either it must admit that morality has jurisdiction over all its actions or it must maintain that it is beyond the jurisdiction of morality: morality demands a total commitment. It seems, then, that the first premise on which the anti-morality view of a government's function is based is valid: no *a priori* guarantee can be given to the effect that in representing moral values in the international field a government will always be acting in ways which further national interest.

To destroy the idea that a government can legitimately have a moral concern internationally, however, the anti-morality argument must also establish the acceptability of its second premise – that a government must pursue the national interest at all costs; only by the establishment of this premise and its conjunction with the first can morality be driven right out of international negotiations. The second premise, however, is not one which commends itself to persons in the liberal-democratic tradition. Whatever may be the *practice* of politicians in our tradition, it is generally felt that a government in pursuing national interest should not abandon the customary morality of diplomatic negotiations. Just as a trustee in pursuing the interests of the principal does not feel justified in breaking the law or departing from accepted moral standards no matter what can be gained by such action, so it is not generally thought to be justifiable for a government to depart from international law or the customary morality of international negotiations in the pursuit of national interest. Politicians in our tradition are no doubt frequently dishonest and hypocritical, as they are in any other political tradition, but we are not grateful to them if our national interest is furthered in ways of which we morally disapprove. In other words, we expect the Government to represent our moral values by enacting its policies, internationally as well as domestically, in a manner which recognizes interests extending beyond its frontiers, shows respect for persons other than its own citizens, and indicates an awareness of the importance of moral ideals. Morality can therefore be allowed to enter the enactment of international policies. And if morality is allowed into the enactment of international policies it may be difficult to say that it has no place at all in the actual policies themselves.

The suggestion that a government may act in the international field in such a way that it represents our moral values as well as our interests is by no means so surprising as it sounds: it is in fact accepted in principle by liberal-democratic countries, and in developing the concept of the moral intermediary we are merely giving shape to ideas which exist in an amorphous form in our present moral outlook. In order to make this claim plausible it is helpful to begin with a case such as a disaster of some kind, say, in Italy or Turkey. Let us suppose that a private person establishes a disaster fund and receives, as is probable, a great deal of support from the people of the nation. In such a case we can say that the person is fulfilling the function of a moral intermediary in the international field, enabling people to carry out their obligations to the wider community of the world at large. In general, we should approve of this function. If the Government is substituted for the private person then it may be said to be fulfilling in a similar manner the function of a moral intermediary. In this capacity the Government is representing the moral values of the people of the country. It is presupposed that most people would feel morally obliged to help if only they could, and the Government is in the position to provide them with the means of fulfilling their moral obligations to humanity at large. The practice of donating public money to international disaster funds is accepted by the public as a whole, and, indeed, many people feel that more ought to be done in this way by the Government. It is arguable in fact that the public responds more freely to appeals to its moral maturity than to national selfishness. If the Government should prove too generous with public money or services to other countries the public can, at least in a liberal-democratic country, express its disapproval in a general election. But there are no instances of a government having lost a general election through being too humanitarian in the international field. On the other hand, support for the agents of repression abroad does meet with moral criticism even when such support appears to be in the national interest.

If what we have said is in general correct then the second premise of the anti-morality argument is not acceptable to the liberal-democratic tradition. Far from believing that the Government should pursue the national interest at all costs, we believe that the Government also has the function of representing our values in the international community. The argument has been that the Government has three functions: it is the ascriptive representative of the people, in that it acts on their behalf as a legally constituted agent; it represents their interests in that, domestically, it is concerned to promote the interest of all the citizens and, internationally, it is similarly engaged in a wider community; it represents the moral values of the citizens, domestically and internationally. Now most often the three functions are compatible, but sometimes it is not possible to represent both interest and moral values and when such conflict occurs priority must be given to moral values. It is for this reason that we have suggested that the Government should be viewed as a moral intermediary, a concept introduced to crystallize attitudes towards government which are not uncommon at the moment. We can therefore reject the objection that a government is failing as a trustee if it does not pursue national interest at all times in the international field.

A second objection to the idea that it is legitimate for a government to have a moral concern in the international field is that whereas there is a stable national society which gives significance to a system of morality within a state, there is no comparable international order to give significance to morality in international affairs. Hence, it is not justifiable for a government to act as if there were: nations are, in Hobbes's terminology, in a 'state of nature' towards each other, and they therefore lack the mutual trust which alone makes adherence to moral values a worthwhile enterprise. This argument rests on two premises, which we shall examine in turn: that no international social order exists, and that the pursuit of morality requires a social order.

It is, of course, easy to provide examples of the behaviour of states towards each other, which suggests that the international order is one of anarchy. But it is equally easy to paint the international scene blacker than it is. Moreover, the hypocrisy of nations can be seen as the tribute which immorality pays to morality. And the very fact that nations feel obliged to provide some sort of moral justification for their international conduct implies a recognition that they are all members of an international community, and they may gradually come to take their responsibilities in the world community more seriously. Kant spoke in terms which implied belief in a world community. For instance, in *Perpetual Peace*, he writes that 'since the narrower or wider community of the peoples of the earth has developed so far that a violation of rights in one place is felt throughout the world, the idea of a law of world citizenship is no high-flown or exaggerated notion. It is a supplement to the unwritten code of the civil and international law . . .'* There is a firm belief in this passage in the existence of an international community and nowadays the belief is much more generally shared. Indeed, a significant difference between Kant's day and our own lies in the fact that what Kant refers to as the 'unwritten code of the civil and international law' has to some extent been written, and the result is a general weakening of the sovereign independence of states. For example, the vast majority of states are prepared to abide by the decisions of the International Court in many matters.

Apart from international law, there exists a large number of institutions which imply the existence of a world community. For example, there is the World Meteorological Organization which both provides benefits for, and requires the co-operation of, every nation in the world. Again, there are a large number of health and medical organizations the services of which are available to all nations, and there are international organizations to deal with transport and communications. Thus, if in Kant's day there was any justification for speaking of a world community and world citizenship, there is much more at the present time. Of course, the organization of the world community must not be thought of as that of a single sovereign state: there is no one centre of power which can enforce international law. The United Nations Organization does not really succeed in this, whatever other merits it may have.

The first premise on which the objection rested was that no world community exists. It has been conceded that no world community exists comparable to the community which exists domestically in any of the long-established nation-states. On the other hand, we have claimed that a system of international law and many other institutions of international organization have developed, especially the U.N. Above all, there is a growing sense of unity and interdependence among the world's nations. The first premise is

* Kant, *Perpetual Peace*, Third Definitive Article for a Perpetual Peace, ed. L. W. Beck, p. 23 (New York, 1957).

therefore at best only partly true, but before assessing its force in its weakened form let us examine the second premise – that morality requires a social order.

It is certainly true that morality, as it was earlier described, becomes both possible and necessary only in a social order of some sort: only in a social order do we find that mutual dependence of persons which is the stuff of morality. To admit this, however, is not to admit that morality requires a highly organized and legally constituted form of order such as we find in an official institution: in fact, morality is more truly itself in a looser form of organization such as we find in a family or a group of friends. It is in groups of this informal kind that most scope can be found for the creative side to morality. The second premise can therefore be accepted, provided we stress that the social order which is necessary for morality need not be tight-knit and formal but may be loose and inchoate.

Applying the discussion of the two premises to the claim that the Government ought to act as a moral intermediary internationally as well as domestically, we may say that the conditions for morality are optimal in international society. Morality is most truly itself in a society of loose and informal organization. But this is also the description which has emerged of international society. Hence, there is limitless scope for the Government to exercise the function of being a moral intermediary. Indeed, morality has really two sides to it: it can maintain and develop an existing social order, and it can create a new social order where one previously existed only in embryo. In other words, if morality is the expression of a social order, it is also the means to one, since it is only as a result of being respected and trusted that people can learn those principles of respect and trust on which the possibility of a social order depend. This point has been brought out by Professor Horsburgh in a discussion of what he calls 'therapeutic trust'.* By 'therapeutic trust' he means a trust which 'aims at increasing the trustworthiness

* H. J. N. Horsburgh, 'The Ethics of Trust', *The Philosophical Quarterly*, 1960.

of the person in whom it is placed'. It would seem to be the case that individual persons can often respond to trust of this nature, and it may be that the official representatives of nations can also do so. If this is the case, then the employment of trust and respect among the ascriptive representatives of the nation may lead to an increase in the trustworthiness of those with whom they are dealing. It may be that the only way to develop our international society is to act as if it existed in a developed form.

4. CONCLUSION

In this chapter we have been concerned with the various questions which arise under the general heading of 'morals and politics'. It is clear that there are a number of different issues to be discussed (and indeed some of them, such as 'private and public morality', have been reserved for a later chapter). In particular, the aim was to elucidate the various criteria we might use in judging the morality of the actions of specific governments, criteria which are often blurred by the use of the umbrella term 'public good'. It also emerged that the question so often asked in the form, 'Should the law enforce morality?', is better framed as, 'What aspects of morality should the law enforce?' And it did seem that a government may legitimately encourage the pursuit of ideals of the good life. Moreover, it may legitimately concern itself with restricting certain forms of vice. The paternalism involved in encouraging ideals and discouraging vice does not raise issues different in principle from legislation, also paternalistic, aimed at the provision of a whole range of welfare services which are now generally expected from the State. Paternalism requires careful interpretation in political terms, however, for while there is something to be said for the claim that a government should preserve the values of an inherited way of life, it does not follow that there should be laws against the wearing of mini-skirts. Finally, we considered how far a government, as a trustee of the national interest, can legitimately be subject to moral rules in the international field, and it emerged that the idea of a government as a representative of moral values – as a moral intermediary – is not entirely alien to liberal ideas.

6

Persons and Roles

1. THE CONCEPT OF ROLE

In the course of the preceding chapters the term 'role' has frequently been used, and it is now time to submit it to analysis.* It deserves detailed analysis for two reasons. In the first place, it is a term which is part of the basic vocabulary of sociologists, anthropologists, social psychologists, political theorists, and other practitioners of the social sciences, but its use in these disciplines is far from clear or consistent. In the second place, an examination of the relationship between role-governed and non-role-governed action brings to light several important problems of social morality, as we shall see in this chapter and the next. Let us begin by considering some of the more common senses of the term.

'Role' is sometimes used in a very wide sense as a class concept. In other words, it is used simply as a way of labelling a group of individuals in virtue of certain properties they have in common. In this very broad sense of the term 'role' we find social scientists speaking of the role of 'football fan', 'cyclist', 'old man', 'invalid', 'bridge player', and so on. This wide sense is not very helpful. It is true that social scientists are concerned with the characteristic patterns of behaviour which people exhibit, but it may be doubted whether they need a technical class concept to refer to these patterns of behaviour. To say that someone is a football fan is already to say that he has certain aims in virtue of which he can be called a football fan. To say that he has the role (in this broad

* My interest in the concept of role was first aroused by Dorothy Emmet, *Function, Purpose and Powers* (London: Macmillan, 1958) and *Rules, Roles and Relation* (London: Macmillan, 1966).

sense) of football fan seems to add nothing by way of description or explanation.

Many social scientists do in fact use the term, secondly, in a more specific way which brings out its dramatic associations. A role was originally a part in a play, particularly a part about which the audience had a set of expectations. For example, the 'types' in morality plays in the later Middle Ages were roles in this sense. Thus, when the devil came on stage the audience had certain standard expectations and hissed at the appropriate places, or again the scolding wife or the jealous husband all had patterns of behaviour which were familiar. There were similar patterns of expected behaviour in Victorian melodrama. Using this analogy, social scientists often speak of someone playing the role of an X or Y. To play the role of invalid is to see oneself as an invalid and expect people to behave to one accordingly. Again, it is common for students to play at being students, for professors to play at being professors and so on.

The same idea is involved in what has come to be called 'projecting an image'. To 'project an image' is self-consciously to adopt a style of behaviour and hope to be identified with this style. 'Style of behaviour' must here be taken to include modes of speech, dress, and general mannerisms. Sometimes these may be modelled on prominent people – 'pop stars', student leaders or the like. In the 1950s, for example, it was common for philosophers – and not just the young of the species either – to model themselves on the alleged mannerisms of Wittgenstein; they saw themselves as intense thinkers and felt obliged to chain-smoke and be rude to one another. At other times the content of the role is made up of some ideal – what an ideal teacher or social worker or army officer ought to be like. The playing of a role in this sense is common in most occupations and social classes. An indication of this is that it is often possible to make a rough guess at people's occupations from the clothes they wear, their type of conversation, and so on. Of course, factors other than 'role-playing' or 'image-projection' are involved here, but it is undoubtedly true that 'role-playing' is a fact of social life, and one to which a social scientist can legitimately pay attention.

It is interesting to note that the evaluation of this phenomenon of social life varies a great deal. One important school of twentieth-century moralists – the Existentialists – devote a great deal of their writings to exploring and condemning the phenomenon. They argue that role-playing represents a kind of insincerity, an evasion of personal relationships and an attempt to pretend to oneself that one's own decisions can somehow be made for one in terms of one's place in the social system. Role-playing is for Sartre, for example, an instance of what he calls 'bad faith' (*mauvaise foi*). In an often-quoted passage from *Being and Nothingness** he describes a waiter who plays at being a waiter. He speaks of the 'grocer's dance, the tailor's, the auctioneer's, by means of which they try to persuade their customers that they are grocers, tailors and auctioneers and nothing more'. Sartre regards this as 'bad faith' because a person is pretending to be nothing except a 'being-for-others', that is, one who merely acts out the role which others have assigned to him, who sees himself as nothing but what people want him to be and who pretends to himself that he is nothing but what his label says he is – a waiter, student, protester or whatever.

It is possible to put the idea of self-conscious role-playing in a more favourable light. One could say that a person might form an ideal of how a follower of a certain trade or profession ought to behave. For example, a teacher might form an ideal of how the true or dedicated teacher ought to behave. This could be based on an actual teacher he had known or an amalgam of several or on a set of qualities. But whatever the causal origins of the ideal one can easily imagine this being done and a person acting as he thinks the ideal teacher ought to act. Now this could be called 'playing the role of teacher', and it seems to be a morally good thing to do and not necessarily to involve any kind of insincerity. It is interesting to note here that linguistic usage can be subtly

* J.-P. Sartre, *Being and Nothingness* (translated by Hazel E. Barnes, London: Methuen, 1957).

directed to impute insincerity when it is believed to be present. Thus we say, 'He was playing the part of the teacher' or 'He was acting the teacher'. These usages bring out the dramatic origins of the concept of role which are dominant in this, the second main sense.

The third sense in which social scientists use the term has less direct connections with the drama, although they are not entirely severed . The third sense is concerned with the function of some activity in a social system. Biological rather than dramatic analogies are the dominant ones here, for this sense in fact involves the characteristic type of explanation to be found in biology – functional explanation. Let us consider the assumptions behind functional explanations in biology.

It is generally said by philosophers of science that there are three assumptions behind the idea of functional explanation in biology. In the first place, there is the assumption that, considered as a whole, the object of study can be regarded as forming a unitary system. But in the second place, this unitary system can be regarded as composed of *elements*. In the third place, the parts or interdependent elements can be said to be causal factors in maintaining the ordered whole in a persisting or enduring state. In the case of biology, where the system is likely to be a living organism, a 'persisting or enduring state' will be analysed in terms of the health or general 'flourishing' of the organism. Where the conceptions of biology are taken over and applied to some artificial system, such as a machine, the third condition can involve the idea of the purpose for which the machine has been made. A standard example of this type of explanation will be sufficient for present purposes. It might be said that the heart beats in order to circulate the blood. It is easy to see from this example how, if we regard the body as the unitary system, all the assumptions are involved, and how, granted the assumptions, the claim is explanatory.

Let us now try to see how functional explanations of this sort can be applied to the study of societies. The first task is to show how the conditions of functional explanation can be met in the study of society. It must be emphasized that this is a large and difficult topic which must here be treated briefly and superficially.

The first assumption requires us to have some rough criterion for the boundary of a society. What are the criteria for the identity of a society? This is not a problem which arises in an acute form in biology since it is usually clear where one animal ends and another begins, and still less is it a problem in the extension of this sort of explanation to artificial systems, since one can always ask the artificer where his machine ends. But the boundaries of a society are not at all easy to draw, especially where they cannot be drawn along nationalistic lines. The solution suggested by the analysis in Chapter 4, which may be adequate for present purposes at least, is that a society ends where the influence of its legal and political institutions ends. This suggestion may well be question-begging for many purposes in sociology, but it will serve in this context. It can also be used in meeting the second and third assumptions of functional explanation. Thus, we can regard society as a differentiated complex in which the elements are interdependent if we think of the institutions of the society as the 'elements', and we can regard these institutions as contributing causally towards the persistence of the society of which they are 'elements'.

Now an institution can legitimately be regarded from two different points of view, that of the sociologist and that of the legal theorist. From the point of view of a sociologist or social anthropologist an institution can be seen as something which is causally operative in maintaining a society in a persisting state. In other words, whatever the specific purposes of the individual persons who operate the institutions, or indeed, whatever the purpose which the institution is commonly thought to perform and (say) receives government money to further, a sociologist or anthropologist may still view the institution in causal terms and may or may not see its existence as functionally explained by the ostensible purpose. For example, an institution such as the House of Lords may be thought to have a certain purpose in the British constitution which could be stated. And there have frequently been arguments about the rights and wrongs of this purpose, whether the institution in its present form is fulfilling the purpose and so on. But equally a sociologist might explain the existence and persistence of the House of Lords in a different explanatory category, by speaking of the effects on society of a device whereby the speed of social change is slowed down and people are given time to adjust to it, learn about its effects, and so on. In other words, an institution can be viewed as a causal factor which brings about certain effects in society regardless of whether the members of the institution have them consciously in mind or not. To concentrate on institutional activity from this point of view is the characteristic concern of the sociologist or anthropologist.

As another example of this first point of view consider what some anthropologists say about a funeral service. They regard it as a way in which people can on some public occasion come to terms with their private grief and indicate to others their new situation in which life must be lived without the deceased. Other anthropologists have discussed the function of feuding in certain societies, and others again have discussed the importance of certain occupations. For example, the blacksmith is thought to have certain magical properties in certain communities.

In all these cases anthropologists and sociologists speak of the 'role' of the institution in question. In using the term in such contexts they have in mind the effects which the institution in fact has in the society, whatever may be the ostensible purpose of the institution. For example, the blacksmith clearly has certain rights and duties as a blacksmith, but these can be distinguished from certain other effects he has on the life of the society. It is in respect of these other effects of which he is the cause that anthropologists may attribute a role to him. In this sense of role even criminals are sometimes said to have a role; they bring about certain effects in society of which they themselves may be quite unaware but which some sociologists believe to be instrumental in preserving the equilibrium of society. Thus criminals may be said to create feelings of hostility in people and hostility towards a common enemy can be a unifying force in a social group. Hence, criminals are said to have a role in society. It may seem strange to regard criminals as having a role, but it is quite a legitimate use of the idea of a functional role; certainly it satisfies the assumptions of functional explanations we have already outlined. Some social scientists claim, indeed, that a social group will generate the functional roles necessary to maintain it in a healthy state. For example, it has been said that a committee will generate provocative members, legalistic members, elder statesmen, and so on. They all have functional roles in the institution in question.

But institutional roles can also be seen from another point of view – that of the political or legal theorist. From this second point of view institutions are regarded as systems of rights and duties which have describable purposes. It is not, of course, denied that they bring about certain effects in the social system, but from the point of view of political or legal theory (or social ethics) some reference must necessarily be made to normative concepts or rules in describing and explaining the operation of the institutions. In other words, from the point of view of sociology and kindred inquiries 'role' is a *de facto* concept and roles are patterns of expected behaviour with certain effects, while from the point of view of social ethics and kindred inquiries 'role' is a *de iure* concept and roles are clusters of rights and duties.

It should be noted that the person who has the role in the sociologist's sense may be quite unaware that he has it, whereas in the sense of the social philosopher the person who has the role must be aware that he has it. For example, whereas in the sociologist's sense it can be said that the criminal has a role, in the social philosopher's sense this would be absurd. Imagine the reaction of judge and jury to the criminal's defence that since he had the role of criminal it was his duty to break and enter to preserve the equilibrium of society! We might say that in the sociologist's sense a person can be said in fact to *have* a role, whereas in the legal or political philosopher's sense he can be said to be *in* a role, or to accept or reject a role. It is the latter sense which is implied by the analysis of authority in Chapter 4. Let us consider this sense in more detail since it is the most important for our purposes and can be used to bring out a number of important problems in social ethics.

A role in the relevant sense, we said, is a cluster of rights and duties with some sort of social function. We can in fact view society (as we indicated in Chapters 2 and 4) as a highly complex set of institutions, each of which consists of one or more clusters of rights and duties. Society from this point of view is a system of legal and political institutions, of banks, factories, schools, trade unions, armed forces, etc. These all give rise to sets of roles. As in the previous, the *causal-functional*, sense, we can regard a role in the normative sense as providing a functional explanation. For example, we can explain what Mr X is doing when he cancels a cheque by referring to his role as banker. He is, of course, bringing about certain effects by his actions, but we cannot adequately describe, far less explain, his action of cancelling the cheque without bringing in normative concepts referring to his rights and duties as a banker. Hence, this sense of role is basic for social ethics.

2. PERSONS AND ROLES

What is the connection between being a moral agent in a social system and having a role in this normative sense?* It might first be suggested that the relationship is one of identity; to be a moral agent is simply to have the rights and duties of some, or a large number, of roles. In terms of this argument morality is seen as a huge system of interlocking roles, and moral agency as a matter of acting *as* a bus conductor, shopkeeper, or professor. Now there is some plausibility in this account of what it is to be a moral agent, but it will not do quite as it stands. An obvious deficiency in it is that it

* This section is based on R. S. Downie, 'Roles and Moral Agency', *Analysis*, December 1968.

cannot account for situations in which a person must choose which role to accept, or in which the requirements of various roles may conflict. Such situations seem to be situations of moral decision (and therefore of moral agency) yet they cannot be explained in terms of the analysis suggested by the first thesis. Again, we can be detached from our roles and laugh at ourselves in them. For example, a teacher speaking sternly to a pupil for an amusing misdemeanour might feel slightly ridiculous as he scolded. This suggests that morality is not exhausted by the idea of a role or set of roles.

A second thesis might be that moral agency is itself a specific role, distinct from that of bus conductor, shopkeeper, etc. Whereas the first thesis reduced moral agency to a set of roles, the second depicts it as one role among others. Now whatever is to be said for or against this thesis it is clear that it is not compatible with the first thesis outlined; for we cannot say both that to be a moral agent is to act as an X or a Y or a Z and that it is to act in a special role distinct from that of an X or a Y or a Z. Hence, we have a second possible thesis about the relationship between roles and moral agency.

The second thesis may be criticized from several points of view. First, it may be said that even if we accept a wide sense of 'role' there must still be some defining properties which would serve to identify it. But what are the rights and duties which would identify the moral role? It is arguable that an answer to this question would either make the moral role too narrow – so as to exclude what are clearly moral decisions or judgements – or too wide to be worth calling a role at all. Secondly, if we accept the second thesis it is difficult to make sense of the point in the first thesis – that to be a moral agent involves acting as an X or Y. It seems that we should need to say that to be a moral agent is to have two roles: the role of shopkeeper, etc. *and* the role of moral agent. But, for a start, this is conceptually uneconomical. Further, let us suppose there is a conflict between the demands of a specific role and those of the moral role. Do the demands of the moral role always take precedence? If they do it is hardly a role in the same sense as the specific roles; if they do not, they are hardly what is ordinarily meant by *moral* demands. Or do we step back, in the case of conflict, into yet another super-moral role? To take this line is to begin a regress.

A third argument against this thesis about moral agency is that if moral agency were itself a specific role then it could be chosen or rejected. But one cannot choose to reject morality, for an apparent choice not to be a moral agent would still be a disguised moral choice; people who say they are rejecting morality are simply rejecting a given morality.

This third argument depends on two premises each of which requires defence. The first is that all roles can be chosen or rejected. Against this it may be contended that some roles – that of 'son', for example – cannot be chosen or rejected. Yet we must avoid confusion over the sense of 'role' which is here relevant; we are considering *social* roles and not biological ones. A man can reject or accept (even if not actually *choose*) the social role of 'son' although the biological one is thrust upon him. Hence the first premise is at least plausible. The second premise is that a man cannot choose not to be a moral agent. Not all philosophers would agree with this premise, but it follows from the general lines of argument in this book. The person who decides he ought not to consider any moral claims on himself, and the person who just does not bother to do so, remain moral agents. They do so in so far as they remain rational members (or indeed beneficiaries) of society. In so far as a person is capable of moral agency (that is, is rational and able to choose), and in so far as by living in a social group he derives benefits, he is in general terms obliged by the rules of the society (whatever he may decide to do about this or that specific rule). In other words, it is not easy to give sense to the idea of choosing not to be a moral agent. Hence, there seem to be three strong arguments against the view that moral agency is itself a specific role.

The view which has been assumed in this essay, and which

follows from Chapter 3, is the trivial one that moral agents are persons, where 'person' means a human being who is capable of exercising rational choice. In other words, there is an irreducibly personal element in moral agency. It should be noted, although the point will not be developed here, that not all human beings are persons in this sense. Thus, human beings who are mentally defective and very young children are certainly *human beings* in the full sense, but they are not fully *persons* in this sense. The point is important for some purposes but it is immaterial in the present context, for we can say either that 'human being' and 'person' mean the same, in which case moral agents are those human beings or persons who can exercise rational choice, or that 'persons' make up a sub-class of human beings, namely those who can exercise rational choice. Whichever way the point is made it is still true that moral agency involves an irreducibly personal element or an element which cannot be reduced to the concepts of role-following.

In this context it may be objected that while 'person' or 'human being' cannot be regarded as an assemblage of roles, 'person' is itself a role-concept. Now there is something to be said for this position. Consider, for example, the history of the term 'person'. Historians of ideas tell us that the term 'person' is derived from the Latin *persona*, which was originally a mask through which came the sound of an actor's voice. The term is then extended to mean a role in the drama or *dramatis persona*, and from there it easily comes to mean a social role. It is in this sense that the term *persona* is used in Roman law, where it stands for someone as a subject of rights. In other words, the history of the word brings out that the term 'person' is not the same as the biological notion of an individual human animal, but is an institutional notion. And we find in Stoic ideas that the notion of a person takes on deeper metaphysical meaning as it deepens in social significance. The Stoic idea of all men as citizens of a single City of God combines the ideas of the supreme metaphysical value of the person with the social idea of the person as the bearer of rights and duties. Hence, in view of this history it might seem natural to regard 'person' as itself a role-concept.

But to say all this is not to say that 'person' simply names yet another role. If 'person' is a role-concept it is not in the same category as other roles; it cannot be, since it cannot be *chosen*. Thus, it is not that Mr X is a committee member, a teacher, a husband, and a person. The connection between personality and roles is different. To bring this out consider the Greek idea of an *ergon*. The Greeks saw the significance of a craftsman as lying in his function and his virtue in being a good craftsman. For each craft or art there was an end and the craft was designed to further the end. This idea was extended in Aristotle to apply to man as such. He was thought to have an *ergon* or function and a particular virtue. This line of thinking suggests that there is a role or function of man or persons as such. Now the point is that there is a category mistake involved here. It is harmless enough to speak of the *ergon* or function or role of a person if by that is meant such things as that persons develop in characteristic ways, do certain things better than they do other things and so on. In short, it is harmless to speak of the role of a person as such if by that is meant only that 'person' is *evaluative* and a different concept from that of 'human being'. But if it means that a given individual human being is an X, Y, Z, *and a person* (where 'person' is put in the same list as the other roles) then the concept of person is distorted, and the concept of role is trivialized. The point of introducing the concept of social role is to stress the often neglected social or impersonal side to morality, to provide a means of conceptualizing the 'what-you-have-got-to-do-as-a-such-and-such'. But if the concept of person is itself analysed in this way there is nothing left to contrast with the impersonal side to morality.

The importance of maintaining a concept of personality irreducible to roles is emphasized, as we have seen, by Existentialist thinking. We have already pointed out that Sartre, for example, condemns the idea that a man is *nothing but* a waiter, soldier, etc. It is true up to a point that for a waiter to be morally good is for him to act as a waiter, and so on for any trade or profession. But the waiter in Sartre's example is being a waiter to the extent that he has forgotten that he is also a person who is a waiter. Again, to see a waiter as *nothing but* a waiter and a shopkeeper as nothing but a shopkeeper is to fail to respect them as persons. Moral agents are always people acting; sometimes they act simply as persons, and sometimes as persons in certain roles or capacities. But however much the rights and duties of the role affect a given action the morality of the action is never wholly reducible to the rights and duties of the role; there is always an irreducibly personal element in any moral action, and a person cannot completely transfer the moral responsibility for what he does to his role.

There are two respects in which a person's responsibility cannot be analysed in terms of the rights and duties of his role. To begin with, it is as a person that he accepts the rights and duties of a given role in the first place. The morality of role-acceptance is therefore necessarily not reducible to that of the role which is accepted. It is as persons plain and simple that we are responsible for the role we accept or reject. Secondly, a person brings to his actions in his chosen role qualities which are not analysable in terms of the role. For example, we may praise a shopkeeper for his courteous and cheerful service, but 'courteous' and 'cheerful' are terms of praise for the person's enactment of his chosen role; they cannot therefore be reduced to the concepts of the role itself.

How far a person can affect the moral quality of his role-enactment will obviously depend on the kind of role in question, for some are much more formal than others. We might in fact compare roles with musical scores. In some scores the composer has indicated not only every note of melody and harmony he wants played but also every nuance of expression he wants brought out. In others, there is simply an unfigured bass line and the performer is left to add his own melody, harmony, and expression. In a similar way some roles leave little scope for individual expression in their enactment (although some people reveal their personality even in the way in which they wield a rubber stamp) while others, such as that of teacher, leave a great deal of scope for individual moral imagination and sensitivity.

In sections 3 and 4 of this chapter (and in the next chapter) I shall illustrate the usefulness of the distinction between impersonal role concepts and personal concepts by applying it to some problems of social morality. In section 3 I shall examine the problem of blending the personal and the impersonal elements in the relationship which exists between a social worker and his client. And in section 4 I shall take two examples of tension between the personal and the impersonal elements in a social situation.

3. AN EXAMPLE: THE SOCIAL WORKER/CLIENT RELATIONSHIP*

It would be generally agreed by experienced social workers that there is a place in the proper approach to a client for both personal and impersonal elements. The degree to which each should be present in a relationship will depend on a range of factors, such as the type of social work – a child-care officer, for example, must surely have a different relationship with a foster child from that of a medical social worker with a temporary adult patient in a hospital – and the age and general temperament of both social worker and client. The dangers of the over-personal and the too-impersonal approaches are well appreciated and stressed in the training of a social worker. There is another danger, however, which is less well appreciated. This danger concerns a confusion which may arise in the understanding of the impersonal attitude, for the term 'impersonal' can characterize at least four different attitudes which may easily be confused. Let us examine these.

The first we shall call a 'cool reactive' attitude. In personal relationships people react in different ways according to temperament and to situation. For instance, in situations of danger normal people react with some sort of fear-response,

* The teacher/pupil relationship provides a comparable example of the blending of personal and impersonal elements.

a response which obviously will vary a great deal according to the nature of the danger and the people concerned. Some people, again, react more strongly than others to situations in which another person is offensive to them. For example, if a shop assistant is rude, or a waiter is needlessly slow, some people make a scene while others remain largely indifferent to the situation. These two examples illustrate the kind of context which might give sense to the idea of a scale of responses to people and situations, and if we accept that there is such a scale we might use the term 'impersonal' of responses at the cool end of the scale. In this sense of the term, an 'impersonal' attitude is not necessarily hostile, but it is detached, indifferent, or not encouraging.

I believe that we may assert that a social worker's attitude ought not to be impersonal in this sense. In other words, it should not be cool, discouraging, or indifferent. True, it ought to be cool in one sense, but it ought not to be 'choking off', but open, receptive, and encouraging.

A second sort of attitude which may be regarded as 'impersonal' is more accurately described as 'official'. Here we are concerned with the attitude of the person who is acting in a role understood as a set of rights and duties. The nature of the attitude which can properly be shown in an official capacity is limited by the rights and duties of the role in which the person is acting. For example, a policeman *qua* policeman may be helpful to a tourist inquiring the way or be courteous even when arresting someone, but clearly the personal attitudes which a policeman may legitimately show in his official capacity are strictly limited. His official attitudes may therefore be described as 'impersonal'.

It will be uncontroversial (I suppose) to say that the social worker's attitude must at least be impersonal in the second sense. Whatever else a modern social worker may or may not be he is at least an official with a set of rights and duties bestowed on him by some welfare institution or the like. The nature of his personal interest in his client will therefore be limited by this official role.

There is a third sort of attitude which might be called

'impersonal', and that is the attitude which ought to be shown by a judge listening to evidence or an examiner conducting an oral examination. 'Impersonal' in this context might also be said to mean 'detached', but it will not mean 'aloof', but rather 'disinterested' or 'unbiased'. (And, of course, 'disinterested' does not mean the same as 'uninterested'.)

Should a social worker's attitude be 'impersonal' in this sense of 'impartial' or 'judicial'? The answer to this will turn on the nature of the social work. For example, the attitudes displayed by workers at a Citizens' Advice Bureau must be impersonal in this third sense. Thus if a client complains about the iniquities of a shopkeeper or landlord, the adviser will listen with sympathy but impartiality, and will, if possible, contact the shopkeeper or landlord to hear the other side of the case. In other words, some judicial impartiality is required, not in the sense that the adviser will make a moral judgement as a result of listening to evidence, but simply in the sense that no solution of the problem can be reached unless both parties in dispute are made aware of how the position is seen from the other point of view. Similarly, a body like a Race Relations Board must be impersonal, in the sense of impartial or judicial, in listening to a complaint and the possible reply to it.

The fourth type of attitude which may be called 'impersonal' is again different in category. It is best illustrated in the attitude which a psychiatrist may have for a patient who is mentally defective or otherwise suffering from some serious disorder. Let us suppose that the patient is abusive to the psychiatrist or offers violence. Now the good psychiatrist will show an attitude to this which can be described as 'impersonal', but what is the nature of the impersonality in this case? Clearly, it is not the judicial impartiality of the third sense; the psychiatrist will not be weighing the pros and cons of the situation and trying to decide whether the abuse or violence was justified. It may be said that the impersonality is of the first, the 'cool reactive' type. Certainly, it may easily be mistaken for such, for it will be cool; but, as we shall see, it will not properly be a reactive attitude at all. It will, of course, be 'impersonal' in the second sense, in that the response will be that of a person acting in a role defined in terms of rights and duties. But the second sense of 'impersonal' does not in this case take us to the nub of the matter. In this case the impersonality is that of the man with a technical problem to cope with. Just as a veterinary surgeon may try to remove a sharp object from an injured animal and be obliged to keep himself clear of the teeth or hooves of the animal, so the psychiatrist in the example will be doing what he can to help the patient but will be obliged to discount any abuse or violence which may be offered. Or, if he does not discount it, he will see it as further relevant data on which he may base his assessments of the patient's problems. In this fourth sense, then, an 'impersonal' attitude is that of the efficient operator who manipulates a given object for one purpose or another and sees nothing personal in any reactions of his object. In the case of the psychiatrist the 'something' or 'object' may also be a human being. The important point may be put in other terms if we say that the object of the impersonal attitude in the fourth sense is seen as being of such a kind that a sufficient causal explanation can be found for its behaviour, whereas the objects of the other three attitudes all describable as 'impersonal' are necessarily persons *acting* in ways open to purposive explanations.

What are we to say of impersonality in the fourth sense? It is certainly an attitude which may be adopted by, say, a psychiatric social worker to some of his particularly difficult cases. He will see them and their behaviour in causal terms. To see them in this way does not, of course, exclude the experiencing of certain emotions towards them, such as pity or sympathy of a sort. But it does exclude regarding them as persons in the full sense. For, without raising again the problem of the freedom of the will, we can maintain that to see human action or speech in this way is to refuse to take it seriously, to refuse to see in it the significance which the agent sees or other people would normally see in it. Clearly, such an attitude lacks the basic respect for personality which is the basis of an acceptable moral outlook. For certain purposes, with certain patients or for part of the time, such an attitude may be legitimate, but my suggestion is that it ought not to be the attitude of the social worker for his normal range of cases.*

In sum, then, the drift of this extended example has been that personal and impersonal attitudes of various kinds should blend in the social worker's dealings with his cases, but impersonality in the fourth sense is a particularly dangerous attitude for a social worker to adopt to the normal range of cases. Let us now turn to what may be called the pathology of persons in roles.

4. PUBLIC AND PRIVATE MORALITY

In considering the relationships between being a person and acting in a role we have so far discussed some of the various ways in which personal qualities or attitudes may *blend* with official rights and duties. In this section I should like to draw attention to quite a different side of the problem and consider some of the ways in which the personal and the official may *pull apart* or otherwise fail to blend. The main issues in this problem are often discussed under the rubric of 'public and private morality' and it may be particularly relevant to the general theme of this book to discuss them in that context.

The problem of the relation between public and private morality really has two sides to it which are not often distinguished. The first of these can be called the problem of private moral views and public duty, and the second the problem of private vice and public duty.†

The first problem is that of analysing the principles which ought to govern a man's conduct in public life when he finds himself in moral disagreement with the policies or particular

* cf. P. F. Strawson, 'Freedom and Resentment', *Proceedings of the British Academy*, 1962, and R. S. Downie, 'Objective and Reactive Attitudes', *Analysis*, December 1966.

† I have discussed these problems in terms of specific examples in *Government Action and Morality*, Chapter IV (London: Macmillan, 1964).

actions he is expected to carry out in his official capacity or role. One extreme view on this we might call the 'ignore-your-own-attitude' view. This is the view that a public servant – politician or civil servant – ought not to have a policy of his own because it will be subversive of good government if public servants expound only those policies with which they personally agree. On this view an official abdicates from responsibility for the actual policies carried out; he becomes simply an implementer.

As against the 'ignore-your-own-attitude' view there is another simple view of the situation, which we may call the 'resign-if-you-disagree' view. This is the view that if a man finds himself in moral disagreement with the duties of his role or the policies he is expected to carry out in it he ought to resign the role. In this view a man is held responsible for the total action which proceeds from the role, as he would have been supposing he had also created both the structure of the role and the policy. In so far as he did not bother to envisage the kind of actions which might be expected of him he is morally blameworthy. Thus, we say, 'You ought to have gone into the implications of your job.' But a job can sometimes turn out to have unexpected features, and then the question of resignation arises. Thus, we say, 'You ought to resign rather than support such policies.' In other words, the argument admits that tensions may sometimes exist between the duties demanded by a person in his social role and his private moral beliefs, and claims that his private moral beliefs must in every case be given priority; he is as fully responsible for the actions he performs in his role as he is for his actions as a private person.

The 'resign-if-you-disagree' view makes an appeal to the liberal cast of mind. The decision to accept a role, on this view, in no way commits one morally to anything one would not otherwise decide to do, for there is always the escape of resignation. In support of this view its exponents might point to Eichmann as one who saw himself simply as an implementer of a policy he was bound to carry out in terms of the duties of his role. Their argument would be that

'I was only obeying orders' is an excuse with a very limited acceptability; the individual ought always to think and decide for himself what he will do.

Now although the 'resign-if-you-disagree' view may sound plausible, there are several factors which may be urged against it, or in qualification of it. The first is that frequent resignations make for unstable government, and this is morally undesirable for the community as a whole. For instance, if we take a disagreement over a minor matter of policy then the 'ignore-your-own-attitude' view is plausible. It is indeed 'subversive of good government' if every politician or administrator has such attachment to his own views that he will refuse to implement those which in any way differ. An official can always offer advice to his superior, and is not morally bound to resign his office if it is not taken, since frequent resignations make for instability in an institution, whether political or commercial.

A second factor which may be urged against the 'resign-if-you-disagree' view is that resignation may have consequences which extend far beyond the particular policy for which the person resigned. For example, a Member of Parliament may decide not to vote against his party even although he disapproves of a particular policy and knows that the constituency he represents disapproves of it: disloyalty to party may on occasion have wider consequences and make for instability.

In the third place, it is important to consider the consequences that resignation will have on the role or policy which is being carried out. A person may hate his role and all its implications but still feel obliged to remain in it because, by opting out, he will leave the role open for someone with no scruples at all. There are various ways in which a role can be enacted, and a person can bring to his actions a quality which may either mitigate or exacerbate its evil effects. Indeed, the actual structure of a role may be gradually altered if it is enacted with a characteristic quality. Just as there is no clear distinction between the interpretation of existing law and the creation of new law, so there is none between enacting an existing role and gradually altering the structure of that role. Even in roles which are basically evil much may be done to mitigate their bad effects by morally enlightened role-enactment.

In view of these morally relevant considerations – the instability which frequent resignations produce, the possible extent of the political repercussions, the possibility of moderating the bad effects of a role by enlightened role-enactment – it is clear that the 'resign-if-you-disagree' view cannot be accepted in its simplicity but must be qualified by the 'ignore-you-own-attitude' view. How far it must be qualified is, of course, a question for the moral agent and depends on the particular situation in which he finds himself. An official might, however, apply three tests before deciding to resign.* He should ask himself whether the issue is one of main principle rather than secondary matters; whether he himself will be directly involved in carrying out the decision; and whether the issue will present a continuing problem in which he will be expected to defend a policy with which he is in fundamental disagreement. Only if the answer to all three questions is in the affirmative should the official resign. Of course, any official might want to add to, or distinguish among, these tests; but clearly the moral situation for an official, or for anyone in a social role, has more complexity than emerges in the simple form of the 'resign-if-you-disagree' view.

It may be objected that to reject the simple form of the resign-if-you-disagree view is also to reject the doctrine, fundamental to the liberal-democratic attitude, that moral responsibility is a function of the adult individual; it is to admit that a person may sometimes perform actions which have moral elements in them for which he is not fully responsible.

But this objection is based on a confusion over the doctrine of individual responsibility. The doctrine is that sane adult individuals (and no others) are to be held totally responsible

* These tests were suggested by Sir Hugh Foot in 'The Principles of a British Public Servant Overseas', *Twentieth Century*, pp. 80–4 (Winter, 1962–3).

for their moral actions: it is not that sane adult individuals are to be held responsible for the total content of their actions. To admit the need for modification in the resign-if-you-disagree view is not to abandon the doctrine that moral responsibility is a function of the individual, but rather to draw attention to the moral problems for the agent which follow from its acceptance. The factors in terms of which I have modified the resign-if-you-disagree view certainly imply that for an official or politician the moral choice must sometimes be a choice between evils. For instance, he may need to choose between the evil of resigning in moral protest, and so leaving the role open for someone with fewer moral scruples, and that of remaining in office, and so implementing a policy of which he does not morally approve. But a choice between evils is still a choice which can be made by a morally responsible person; no moral abdication is involved. After all, the moral situation of the private person not acting in a social role may occasionally be such that he must choose between evils, and we should not on that account alone deny his moral responsibility. Indeed, it may be said that the agent is never responsible for the total content of an action. The situation in which he finds himself is never one of which he is sole creator, since he is born into an environment not of his making; his moral actions, being reactions to his situation, are necessarily affected by it.

Let us turn now to the second problem of private and public morality – that of whether a person is unfitted for holding a public office by a private moral life which many in the community would condemn. Again we find that there are two views commonly expressed on this. The first we may call the 'continuity' view. It is to the effect that there is a continuity or solidarity between the moral actions which a person carries out as a person and those he carries out as an official or politician. Hence, if he lives a life of sexual vice this gives good grounds for sacking him as a politician or official. This view is frequently expressed in the newspapers.

It is obvious that as an unrestricted thesis the continuity view is not plausible. A minister may be unnecessarily unpunctual at meal times or mean with tips, but these failings, although they may be matters of private morals, hardly affect the public life of an official or politician. If the continuity view is to be plausible, therefore, we must find some way of distinguishing those aspects of private morals which are relevant to public life from those which are not. A way of drawing this distinction may be in terms of the idea (already introduced in section 2) of the 'carry over' of certain personal moral qualities from private to public life. Thus many moral virtues, such as honesty, are obviously relevant to public life, since a failure in honesty in private life gives grounds for inferring a probable failure of a similar kind in public life. In its modified form, however, the continuity view would on first sight exclude as irrelevant the sexual morality of politicians, and on that account it would not be acceptable to some people. But since there is no obvious connection between sexual morality and political reliability the onus of proof seems to rest on those who assert that there is a connection. Two main arguments (apart from those with premises derived from religious belief) may be used to assert a general relevance of sexual morality to political life.

The first is the argument from national security: a public figure may have his secrets wheedled out of him by his mistress. In this form the security argument is not convincing. Provided the official exercises discretion in his choice of mistresses no danger to security need arise. There is, however, a stronger form of the security argument to the effect that the sexual failings of a public figure may render him liable to blackmail. The force of blackmail in this context will depend on the domestic situation of the public figure, the state of public opinion on sexual behaviour and the attitude of the Press, and, while blackmail may in isolated cases be applied to a politician, it remains true that many major politicians whose sexual morals would not bear much scrutiny have survived these dangers unscathed. It is worth noting, however, that homosexual practices make a public figure much more liable to blackmail. But even on this issue public opinion is becoming more liberal. At best, then, the security argument has force only in specific cases and does not establish a general connection between sexual behaviour and political reliability.

The second argument for the relevance of sexual morality to political morality is the argument from the importance of setting a good public example. The force of this argument, however, depends entirely on the suppressed premise that the sexual morals of public figures must be exposed to public scrutiny, and this is a premise of doubtful acceptability in liberal-democratic morality. This is especially true in view of the methods and procedures which are involved in exposing a person's private life to public scrutiny: rumour, innuendo, and the intrusion into private life by reporters and photographers. It is true that the Press defends its activities in this context by referring to national security, but this may often be merely a pretext for the exploitation of sexual scandal to satisfy the prurient curiosity of newspaper readers. In so far, then, as the public example argument implies a context of Press intrusion and gossip, it is not compatible with the principle of respect for persons which is at the root of our moral outlook.

It therefore seems doubtful whether there is a satisfactory argument which establishes the general relevance of private sexual morals to public life, although in specific cases and for specific reasons it may be justifiable to caution or sack a politician for his private sexual morals. It would indeed be surprising if there were a general relevance of sexual morals to public office in view of the wide divergencies in attitude towards the former among the distinguished public figures in the history of politics. The continuity view of the relation between private morals and public office can certainly be said to improve on the separate compartment view, to the extent that there are some moral virtues and vices which carry over easily from private life to public life, but it must be stressed that moral judgement on these virtues and vices is pertinent only in so far as the role-enactment of the public figure is affected.

5. CONCLUSION

In this chapter the concept of social role, which has been used extensively in this essay, has been examined and its usefulness for social ethics illustrated. Among the many senses which the term has in the writings of social scientists the one which is most relevant for our purposes is that of role as a set of rights and duties. Moral behaviour from the point of view of social ethics can be seen as a matter of persons acting in roles in this sense. When a person acts in a role in this sense his action can be said to have both a personal and an impersonal side. The term 'impersonal' is ambiguous, however, and we looked at four possible meanings it may have. The conceptual distinctions thus uncovered were found to be helpful in analysing the nature of the relationship which a social worker may have with his client. In the fourth section of the Chapter I considered what might be called the 'pathology' of role-governed action – where the personal and the role elements are in various respects in disunity. The problems we have raised are all practical ones and the answers to them must therefore in the end consist in the formulating of practical policies or the making of decisions by the persons or officials concerned. I hope, however, that the discussion has shown that certain conceptual distinctions which philosophical analysis can enable us to draw are relevant to the understanding of the problems.

7

Resenting, Forgiving, Punishing, and Pardoning

I. RESENTING AND FORGIVING

The distinction between what we can do as persons and what we can do as persons-in-social-roles is all-important for social ethics. In this chapter we shall continue to explore features of the distinction by considering a group of related concepts: resenting, forgiving, punishing, and pardoning. Let us begin with the concepts of resenting and forgiving, for, despite differences between them, there are important similarities in their logic which contrast with similarities in the logic of the second pair of concepts.

In the first place, we can say that consciousness of injury to oneself is a logically necessary condition of both resentment and forgiveness. In other words, if A can legitimately be said to feel resentment towards B, or to forgive B, then A logically must be conscious of having been injured by B. Moreover, for the forgiveness or resentment to be logically appropriate it must be the case that the injury is believed to have been intentionally or at least negligently inflicted. It is not necessary, however, that B should realize that he is injuring A for A's resentment or forgiveness to make sense; B may simply be pursuing his own interests at all costs and be unaware that he is injuring A. But provided that a reasonable person would have realized that the action constituted an injury to A we should regard A's resentment or forgiveness as appropriate. After all, people may be forgiven although they 'know not what they do'. With these qualifications, then, we shall say that consciousness of injury is a logically necessary condition of both resentment and forgiveness. What further constitutes resentment and forgiveness?

The answer is not difficult in the case of resentment. If A resents what B has done to him he will act towards B, at least temporarily, in a resentful manner. In other words, he will feel badly done by and will think hostile thoughts and display hostility in his behaviour. There is no one sort of mental state or behaviour which is resentment – human beings are remarkably ingenious in this respect – but whatever specific form it takes it will be one kind of reaction, a hostile one, to the consciousness of injury.

The other kind of reaction to such consciousness is to forgive.* What is it to forgive? It might at first be described as excusing, overlooking, treating indulgently or, in a word, as condoning. 'Forget it,' we say, 'it was nothing.' Such a mode of behaviour, however, might be considered morally wrong in several different ways. In the first place, to minimize the injury where the injury has been severe – to say, 'It does not matter', when it does – is less than honest. Moreover, insincerity of this kind may give rise to a secret sense of grievance, which is the antithesis of the forgiving spirit. In the second place, our reaction to the offender may be governed by an understanding of the person and the causes or reasons which resulted in the injury, and we may see the person as the victim of psychological pressures he cannot resist. For instance, we may forgive because we come to understand that the person who injured us was an alcoholic. Now in such circumstances it may well be morally wrong – and it would certainly be imprudent – to condone the injury by overlooking it or by treating the person concerned indulgently, as if he had not inflicted the injury. On the contrary, forgiveness in this context may require stern treatment of him, with exposure and investigation of his weakness. In the third place, injury to a person also involves a moral wrong, and, however much the forgiver may be prepared to play down the injury he has received, it is not morally

* I have discussed forgiveness at more length in *Philosophical Quarterly*, April 1965.

appropriate that he should also disguise the fact that a moral wrong has been committed. In other words, although A may be prepared to overlook the injury that he has received, his attitude may be judged morally inappropriate in so far as it plays down the moral wrong of treating people as he has been treated. If, then, the attitude of condonation is sometimes morally inappropriate, it cannot be the attitude of forgiveness, because readiness to forgive is a virtue and the exercise of a virtue is never morally inappropriate. Is there any other way of characterizing what it is to forgive?

It is not satisfactory to say that the mere uttering of the words 'I forgive you' constitutes forgiveness. The uttering of these words, or their equivalent, is certainly not sufficient to constitute forgiveness. Unless the words are accompanied by the appropriate behaviour we shall say that A has not really forgiven B. In this respect forgiving differs from promising. Whereas the uttering of 'I promise' does, at least in most circumstances, constitute a promise even although the appropriate behaviour is not forthcoming, the uttering of 'I forgive' does not constitute forgiveness unless the appropriate behaviour is forthcoming. It is true that forgiving is like promising in that to say, 'I forgive you', is to raise certain expectations which may or may not be fulfilled. But if the expectations are not fulfilled in the case of promising it is still true that a promise has been given, although a false one, whereas if they are not fulfilled in the case of forgiving we do not allow that there has been forgiveness at all. The uttering of a formula is therefore not a sufficient condition of forgiveness. It is doubtful, moreover, whether it is even a necessary condition. We can forgive a person without his knowledge or in his absence, merely by altering our attitudes and behaviour towards him.

But in what way must we alter our behaviour or attitudes for them to satisfy the conditions for forgiveness? The answer may seem hard to obtain if we look for some *special* mode of behaviour. The truth is rather that the forgiving spirit is simply the mode of behaviour which is always appropriate to interpersonal behaviour – that of respect or *agape*, or a sympathetic concern for the dignity of persons conceived as ends in themselves. An injury involves the severing of the relationship and forgiveness its restoration. At times, where the injury is trivial, this may involve merely condoning or letting off lightly, but most often such an attitude is not consonant with the attitude of respect. Respect involves the treatment of other people not just as sentient beings but as beings who are rational and able to obey moral rules and pursue moral values just as the forgiver himself can. The forgiver is required to prevent any barrier remaining permanently between him and the forgivee and to renew trust in him. It is the exhibition of this attitude in action which, together with a belief that injury has been sustained, constitutes forgiveness.

We have seen that the forgiving spirit is not in fact different from the attitude of respect which should always characterize interpersonal behaviour, but this attitude becomes the forgiving spirit in a context of injury. This fact enables us to understand why it is possible to speak of a forgiving disposition. Some people, by nature, grace or moral achievement, exhibit more or less permanently the spirit of respect in their behaviour. For them, the sustaining of injury does not create a break in the attitude: it can be said, almost without paradox, that they have forgiven before they have been injured. For most people, however, to receive a serious injury is to have the outgoing and positive attitude of respect broken or at least disturbed. Forgiveness is here an episode which involves the conscious removal of the barrier which injury has raised.

2. THE JUSTIFICATION OF RESENTMENT AND FORGIVENESS

Let us now consider how the concepts of resentment and forgiveness can be morally justified. Resentment as such probably cannot be. If A is injured by B then he may well react by showing resentment. This is understandable and is not morally condemned. But if A *continues* to show resentment then he will be condemned in many moral communities. It is, of course, possible for him to continue to condemn, deplore, and disapprove of B, but that is different from saying that he may continue to show resentment. Resentment is a natural (as opposed both to pathological and to artificial) response to the consciousness of injury, but it ought nevertheless to be replaced by forgiveness. This would be a widely accepted moral view.

Is it possible to justify the assumption that readiness to forgive is a virtue and unwillingness to try to forgive (or continuing to resent) a vice? In so far as justification is possible at all it is only within moral systems which share a certain outlook on life. The justification, that is, will not convince anyone influenced by such an ethic as that of Nietzsche, who would regard readiness to forgive as a characteristic of 'slave morality'. If it is objected that, being limited in this way, what is provided cannot be justification at all, we can retract the term 'justification' and admit that all we are doing is to point to the position of forgiveness in moral systems, religious or secular, influenced by Judaistic conceptions.

Even within these systems, however, there is a provisional difficulty which must be met in any claim that readiness to forgive is a virtue. The difficulty arises over the fact that forgiveness implies consciousness of injury, and readiness to forgive may therefore suggest readiness to feel injured. But there is something morally offensive about undue sensitivity to injury, even although forgiveness follows from consciousness of injury. It may seem, therefore, that, even within a moral system friendly to forgiveness, readiness to forgive cannot be a virtue, because a virtue is not good in some contexts and bad in others, but is always good. The solution of the problem lies in pointing out that it is not readiness to forgive which constitutes the moral offence in the situation described: it is undue sensitivity to injury which is morally offensive. After all, undue sensitivity to injury coupled with *unreadiness* to forgive creates a much worse situation than the one raised in the objection. We can therefore regard the assumption that it is always good to forgive as being undisturbed by the fact that undue sensitivity to injury is less than admirable. The problem remains, however, of showing the place of the assumption within moral systems influenced by Judaistic conceptions.

As a Christian virtue, forgiveness must be justified by essentially Christian conceptions. (This is not, of course, to say that Christians would necessarily deny the legitimacy of other justifications.) The Christian justification would presumably run like this: 'Since your Heavenly Father has forgiven you, you ought also to forgive others.' In other words, forgiveness is justified as being a response to injury fitting in creatures who are themselves liable to error.

In the secular morality of the West the fundamental principle in terms of which more specific moral rules and virtues require to be justified is that of respect for persons as ends in themselves. It is, however, easy to justify the forgiving spirit in terms of this principle because the attitude of respect which constitutes the forgiving spirit is that very principle of respect for persons in its practical application.

3. PARDONING AND PUNISHING

Let us now compare the logic of the second pair of concepts we are considering – pardoning and punishing – with that of forgiving and resenting. The first difference between the concept of forgiveness and that of pardon emerges if we ask what constitutes pardoning. To pardon a person – whether this be done by monarch or club committee – is to let him off the merited consequences of his actions; it is to overlook what he has done and to treat him with indulgence. To pardon is in fact to condone, but we have seen that, whatever it may be, to forgive is not to condone. If we ask, in the second place, for what a person is pardoned, we find that it is necessarily for an offence (i.e. a violation of rules), whereas it is necessarily for injuring (i.e. hurting, physically or psychologically) that a person is forgiven. If it is possible at all to speak of pardoning an injury it is because the injury is seen as an offence against some normative order. A monarch, for example, pardons offences against the law of the land, and while he may (or may not) pardon injuries against his own person, he does so in so far as injuries to the monarch are offences against the embodied law of the land. Similarly, in pardoning an erring member, a club committee is concerned with offences against its rules rather than with injuries sustained by its members. It is logically impossible, on the other hand, to forgive an offence, except in so far as an offence is conceived as an injury. Now, if it is necessarily an offence which is condoned in a pardon, pardoning must possess another feature which distinguishes it from forgiving: since an offence is a violation of a normative order only someone who is rightfully appointed or formally constituted according to the secondary rules is qualified to condone such a breach. While anyone may condone an offence in the sense of regarding it with indulgence, it is only the monarch or the club committee who can pardon it, and they do so by considering not how it affects them personally as injury but how it bears on the rules. But there is no such formal restriction on the forgiver: anyone who has been injured is qualified to forgive his injurer. The crucial difference, then, between pardoning and forgiving is that we pardon as officials in social roles but forgive as persons.

It is now possible to explain why 'I pardon you' can be what is called a performative utterance whereas 'I forgive you' cannot.* The first utterance can be a performative because, when uttered by the appropriate person in the appropriate context, it constitutes a pardon. When the monarch says, 'I pardon you', he is in fact pardoning the offender. In other words, by uttering the formula he sets in motion the normative machinery whereby the offence will be overlooked: he himself need do no more in his official capacity. To say, 'I forgive you', however, is not in a similar way to set anything

* For an analysis of the term 'performative utterance' see J. L. Austin, 'Performative Utterances', *Philosophical Papers* (Oxford: Clarendon Press, 1961).

in motion. The forgiver is merely signalling that he has the appropriate attitude and that the person being forgiven can expect the appropriate behaviour.

We can now see from this analysis the importance for social ethics of distinguishing the two groups of concepts we identified in our previous chapter. One group includes those concepts, such as forgiveness and resentment, which refer to actions we can perform or dispositions we can manifest as persons: the other includes pardon and punishment, which refer to actions we can perform only in social roles.

As a further illustration of the usefulness of this distinction let us (by way of digression) consider its applicability to a problem in the philosophy of religion – that of the sense in which only God can be said to forgive sins. If 'sin' is defined as 'injury to God', it is true, and indeed trivially true, that only God can forgive sins. It is possible, however, that something more may be meant by 'sin' than 'injury to God', and that this something more may be described as an offence against an order of moral values. If we assume that this is what is meant (and that the conception can be made intelligible), then some meaning other than a trivial one can be attached to the claim that God and only He can forgive sins. The truth expressed in this claim, however, is not concerned with forgiveness but with pardon. If we say that God embodies impersonal moral values in a manner similar to that in which a monarch embodies the laws of the land or a father embodies the rules of the household, we can say that God is able to pardon offences in the sense that He can overlook them and treat the offenders with indulgence. Failure to distinguish such pardoning from forgiving may lead to conceptual muddles in theology. Let us return to social ethics, however, and develop the distinction between what we can do as persons and what we can do as officials in social roles by contrasting the concepts of resentment and punishment.

It might be objected here that resenting, unlike punishing, is not something we can be said to *do* at all. But the required distinction can be brought out if we consider the natural expression of resentment in action, namely, revenge. Revenge is like punishment in that it involves, firstly, the infliction of pain of some kind on the believed injurer. In the second place, revenge is like punishment in that both must be intentionally directed towards the person concerned. If the injurer or criminal harms himself by accident in the furtherance of his misdemeanour then this is no doubt gratifying to the injured party ('It serves you right') but it does not constitute punishment or revenge, unless there is a belief that God has arranged the accident. But despite these two similarities between revenge and punishment the differences are more important. There are three main differences.

In the first place, whereas revenge (or resentment) must logically be for an actual or imagined *injury*, punishment must logically be for an actual or imagined *offence*. This is the point we have already encountered in contrasting forgiveness and pardon. In other words the point is that punishing and pardoning necessarily involve the idea of rules, of a normative order which has been violated, whereas resenting and forgiving involve only the idea of believed harm or injury. It follows from this, secondly, that whereas resentment, revenge, and forgiveness are directed towards a believed *injurer* for the injury he has inflicted, punishment, like pardon, is directed towards a supposed *offender* for his offence. In the third place, whereas resentment and forgiveness emanate from the people who have been injured (or from those with whom they are in some way identified) punishment and pardon are precisely not administered by injured parties, but rather by an official whose role is constituted by the same normative order which set up the rules which have been violated by the offender.

Against the third point it might be argued that if a school teacher is kicked in the shins by a pupil whom he proceeds to punish, then we have a case of punishment being carried out by the injured party. The reply to this objection is to distinguish the two capacities in which the teacher is acting. When he is assaulted by the pupil he sustains an injury which he may well resent and which may tempt him to retaliate. But if he does retaliate or show resentment as a person, he may be acting in a morally improper way. To say this, however, is not to say that he ought not to punish the pupil, but rather that if he does decide to punish he is acting as an authorized officer of his institution against someone who has broken its rules. It would be logically possible for the punishment to be administered by another teacher, and this in fact is done in some schools. It is equally possible logically (and even psychologically, despite the dangers of self-deception) for a teacher to punish a pupil as an offender against the rules while at the same time forgiving the pupil for the personal injury he has received. Thus punishing and forgiving are quite compatible in that they are concepts from different logical orders.

4. THE JUSTIFICATION OF PUNISHMENT

In the case of forgiveness it was not difficult to show how the practice can be justified, at least within one type of moral view, for to forgive is simply to restore the morally proper interpersonal attitude which has been upset by an injury. But punishment does not seem to involve the restoration of normal interpersonal behaviour and attitudes. On the contrary, it seems nearer resentment, which we argued is morally unjustified if it is sustained. Indeed, some have said that punishment is simply institutionalized revenge. Hence, the justification of punishment as a practice is more complex and controversial than the justification of forgiveness.

Traditionally two very general sorts of justification have been offered for the practice of punishment: retributivist and utilitarian. According to the theory of retribution, punishment is justified in so far as it is a morally fitting response to the violation of a law. Sometimes the theory is expressed in the ambiguous form: punishment is justified by guilt. This form of words is ambiguous because it may mean that guilt is a necessary condition of punishment, or it may mean that guilt is a necessary and a sufficient condition of punishment. The first of these claims is entirely innocuous – it is indeed part of the definition of punishment that it is for an offence (real or supposed) – and can be conceded by all without further ado. But in so far as the first claim simply states part of the definition of punishment it cannot constitute any part of its justification. Hence, in so far as the retributive theory sets out to be justificatory it logically must involve the more radical of the two claims – that guilt is a necessary and a sufficient condition of the infliction of punishment on an offender. Is the theory plausible?

It might be said against it that it does not reflect actual practice either in the law or in more informal normative orders such as schools or the family. In these institutions we find such practices as the relaxation of punishment for first offenders, warnings, pardons, and so on. Hence, it may seem that actual penal practice is at variance with the retributive theory if it is saying that guilt is a necessary and a sufficient condition of punishment. Only in games, it may be said, do we find a sphere where the infringement of rules is a necessary and a sufficient condition of the infliction of a penalty, and even in games there may be a 'playing the advantage' rule.

It is not clear, however, that the retributivist need be disturbed by this objection. He might take a high-handed line with the objection and say that if it is the case that actual penal practice does not reflect his theory then so much the (morally) worse for actual practice. Whatever people in fact do, he might argue, it is morally fitting that guilt be a necessary and a sufficient condition of the infliction of punishment.

A less high-handed line, and one more likely to conciliate the opponents of the retributivist theory, is to say that the statement of the theory so far provided is to be taken as a very general one. When it is said that guilt is a necessary and a sufficient condition of punishment it is not intended that punishment should in every case follow inexorably. There are cases where extenuating circumstances may be discovered and in such cases it would be legitimate to recommend mercy or even to issue a pardon. It is unfair to the retributivist to depict his theory as the inflexible application of rules. There is nothing in the theory which forbids it the use of all the devices in the law and in less formal institutions whereby punishment may in certain circumstances be mitigated. Even the sternest of retributivists can allow for the concept of mercy. Perhaps the objection may be avoided completely if the theory is stated more carefully as: guilt and nothing other than guilt may justify the infliction of punishment. To state the theory in this way is to enable it to accommodate the complex operation of extenuating factors which modify the execution of actual systems of law. But to say this is not yet to explain why guilt and nothing other than guilt may justify the infliction of punishment. Indeed, we might ask the retributivist to tell us why we should punish anyone at all.

Retributivists give a variety of answers to this question. One is that punishment annuls the evil which the offender has created. It is not easy to make sense of this claim. No amount of punishment can undo an offence that has been committed: what has been has been. Sometimes metaphors of punishment washing away sin are used to explain how punishment can annul evil. But such ideas can mislead. Certainly, the infliction of punishment may, as a matter of psychological fact, remove some people's *feelings* of guilt. But we may query whether this is necessarily a good thing. Whether or not it is, however, it is not the same as annulling the evil committed.

A second retributivist idea is that the offender has had some sort of illicit pleasure and that the infliction of pain will redress the moral balance. People speak of the criminal as 'paying for what he has done' or as 'reaping a harvest of bitterness'; the infliction of suffering is regarded as a fitting response to crime. But this view may be based on a confusion of the idea that 'the punishment must fit the crime', which is acceptable if it means only that punishment ought to be proportionate, with the idea that punishment is a fitting response to crime, which is not so obviously acceptable. It may be that the traditional *lex talionis* is based on a confusion of the two ideas. At any rate, it is the idea of punishment as the 'fitting response' which is essential to the retributivist position.

A third claim sometimes made by the retributivists is that an offender has a right to punishment. This claim is often mocked by the critics of retributivism on the grounds that it is an odd right that would gladly be waived by the holder! It might be thought that the view can be defended on the grounds that the criminal's right to punishment is merely the right correlative to the authority's duty to punish him. Such a defence is not plausible, however, for, even supposing there is a right correlative to a legal authority's duty to punish criminals, it is much more plausible to attribute this right to the society which is protected by the deterrent effect of punishment.

There are two arguments, more convincing than the first, which can be put forward in defence of the view that the offender has a right to punishment. One of these requires us to take the view in conjunction with the premise that punishment annuls evil. If punishment can somehow wash away the guilt of the offender then it is plausible to say that the offender has a right to be punished. The difficulty with this argument, however, is that it simply transfers the problems. It is now intelligible to say that the offender has a right to punishment, but, as we have already seen, it is not at all clear what is means to say that punishment annuls evil.

The other argument invites us to consider what the alternative is to punishing the criminal. The alternative, as we shall shortly see, may be to 'treat' him and attempt a 'cure' by means of various psychological techniques. Now this would be rejected as morally repugnant by many retributivists, on the grounds that an offender has freely decided to break the law and should be regarded as a self-determining rational being who knew what he was doing. An offender, so described, may be said to have a right to *punishment* (as distinct from psychological treatment, moral indoctrination, or brain-washing in the interests of the State). Such an argument produces a favourable response from many criminals who, when they have served their time in prison, feel that the matter is then over and that they ought to be protected from the attentions of moral doctors and the like. So stated, there may be some truth in the 'right to punishment' doctrine, however easy it is for the sophisticated to mock it.

So far we have been concerned to suggest detailed criticism of the retributivist theory. But there are two general criticisms which are commonly made of it at the moment (for it is a most unfashionable theory in philosophical and other circles). The first is that in so far as its claims are intelligible and prima facie acceptable, they are disguised utilitarian claims.

This criticism is valid against certain ways in which retributivists sometimes state their claims. For example, they have sometimes regarded the infliction of punishment as the 'emphatic denunciation by the community of a crime'. But we might well ask why society should bother to denounce crime unless it hopes by that means to do some good and diminish it. A second example of a retributivist claim which easily lends itself to utilitarian interpretation is that the infliction of punishment reforms the criminal by shocking him into a full awareness of his moral turpitude. The question here is not whether this claim is in fact plausible (criminologists do not find it so) but whether the justification of punishment it offers is essentially retributivist. It seems rather to be utilitarian, and in this respect like the more familiar version of the reform theory we shall shortly examine.

It seems, then, that this criticism does have some force against certain retributivist claims, or against certain ways of stating retributivist claims. But the criticism does not do radical damage to retributivism. Provided the theory is stated in such a way that it is clear that the justification of punishment is necessarily only that it is a fitting response to an offender, then the fact that punishment may sometimes also do good need not count against retributivism. It is, however, an implication of retributivism that it must sometimes be obligatory to punish when punishment is not expected to do any good beyond itself.

It is precisely this implication which is used as the basis for the second criticism of retributivism. This criticism is a straightforward moral judgement, that the theory is morally objectionable in that it requires us to inflict punishment, which is by definition unpleasant and therefore as such evil, for no compensating greater good. The critics therefore invite us to reject the theory as a barbarous residue of old moral ideas.

In the context of philosophical analysis it is important to avoid taking sides in moral argument, as far as this can be done. But it may be worth pointing out that perhaps the great majority of ordinary people have sympathy with some form of retributivism. Contemporary philosophers often appeal to what the 'ordinary moral agent' would do or say in certain circumstances, or what the morality of 'common sense' would hold on certain topics, but if they make this appeal in settling questions of the justification of punishment they may find that the ordinary person adheres to some form of retributivism. Since there is generally something to be said for an appeal in moral matters to what people ordinarily think, let us consider whether the implication of retributivism is really so morally repugnant.

The implication is that on some occasions it will be obligatory to punish an offender even although this will do no good beyond itself, and that even when some further good is in fact accomplished by the punishment this is irrelevant to its justification. Now it may be that this implication is thought to be morally objectionable because the clause 'no good beyond itself' is equated with 'no good at all'. But a retributivist will claim that the mere fact of inflicting punishment on an offender is good in itself. It is not that a *further* good will result (although it may do so) but that the very fact of the punishment is fitting and to that extent good.

This argument may be made more convincing by an example. Let us suppose that a Nazi war criminal responsible for the cruel torture and deaths of many innocent people has taken refuge in South America where he is living *incognito*. Let us suppose that he has become a useful and prosperous member of the community in which he is living. Let us suppose that the whereabouts of this criminal are discovered and it becomes possible to bring him to justice and punish him. It is very doubtful whether such punishment will have any utilitarian value at all; the criminal will not in any way be 'reformed' by treatment, and the deterrence of other war criminals seems an unrealistic aim. We can at least imagine that the punishment of such a criminal will have minimal utilitarian value and may even have disvalue in utilitarian terms. Nevertheless, many people might still feel that the criminal ought to be brought to justice and punished, that irrespective of any further good which may or may not result from the punishment, it is in itself good that such a man should be punished for his crimes. To see some force in this special pleading is to see that retributivism is not completely without moral justification although it can never on its own constitute a complete theory of the justification of punishment.

I have tried to find merits in the retributivist theory because it is frequently dismissed with contempt by philosophers and others at the present time, but we shall now consider theories of the justification of punishment with more obvious appeal. These are all different forms of utilitarianism and they therefore have in common the claim that the justification of punishment necessarily rests in the value of its consequences. There are two common forms of utilitarian justification: in terms of deterrence and in terms of reform.

According to the deterrence form of the utilitarian theory the justification of punishment lies in the fact that the threats of the criminal law will deter potential wrong-doers. But, since threats are not efficacious unless they are carried out, proved wrong-doers must in fact have unpleasant consequences visited on them. The increase in the pain of the criminal, however, is balanced by the increase in the happiness of society, where crime has been checked. The theory is modified to account for two classes of people for whom the threats of the law cannot operate: the cases of infants and madmen on the one hand, and on the other hand cases in which accident, coercion and other 'excusing conditions' affected the action. In such cases the threats of the law would clearly have little effect and punishment would therefore do no social good.

A deterrent theory of this general sort has often been accepted by utilitarian philosophers from the time of Bentham, but it is frequently criticized. The most common criticism is that it does not rule out the infliction of 'punishment' or suffering on the innocent. Utilitarians, that is, argue that where excusing conditions exist punishment would be wasted or would be socially useless. But their argument shows only that the threat of punishment would not be effective in particular cases where there are excusing conditions; the infliction of 'punishment' in such cases would still have deterrent values on *others* who might be tempted to break the law. Moreover, people who have committed a crime may hope to escape by pleading excusing conditions and hence there would be social efficacy in punishing those with excuses. Presumably this is the point of 'strict liability' in the civil law. But if the deterrent theory commits us to such implications it is at variance with our ordinary views, for we do not accept that the punishment of the lunatic (say) is permissible whatever its social utility.

Utilitarians have sometimes tried to meet this objection by arguing that a system of laws which did not provide for excusing conditions might cause great misery to society. There would be widespread alarm in any society in which no excusing conditions were allowed to affect judicial decisions in criminal cases, and indeed (a utilitarian might argue) such a system might not receive the co-operation of society at large, without which no judicial system can long operate. The utilitarian reply, then, is that while punishment is justified by its deterrent value we must allow excusing conditions since they also have social utility.

It is doubtful, however, whether this reply is adequate. For if excusing conditions are allowed only in so far as they have social utility there remains the possibility that some unusual cases may crop up in which the infliction of suffering or 'punishment' on an innocent person would have social utility which would far outweigh the social utility of excusing him. This is not only a logical possibility on the deterrent theory but a very real possibility in communities in which the 'framing' of an innocent person might prevent rioting of a racial or religious kind. But this implication of the deterrent theory is at odds with widely accepted moral views. Rather it would be held that the rights of the individual must come before the good of society. This does not mean that the rights of the individual must never be sacrificed to the general good but only that there are certain basic rights, the rights of man, or rights which belong to persons as such, which must never be violated no matter what social good will accrue. It is the weakness of the deterrent version of the utilitarian justification of punishment that it cannot accommodate this truth. Here the deterrent theory contrasts adversely with the retributive theory which does stress the rights of the individual against those of society.

Despite this criticism, however, the deterrent theory cannot be entirely dismissed for it does have relevance at the level of legislation. Whereas it is a failure if it is regarded as an attempt to provide a complete justification of punishment it does succeed in bringing out one of the functions of punishment – that of acting as a deterrent to the potential criminal.

The second conception in terms of which utilitarians try to justify the infliction of punishment is that of 'reform'. They argue that when a criminal is in prison or in some other detention centre a unique opportunity is created for equipping him with a socially desirable set of skills and attitudes. The claim is that such a procedure will have a social utility which outweighs that of conventional punishment.* The

* This theory must be distinguished from that mentioned in the discussion of retributivism – that conventional punishment reforms by 'shocking' the criminal into an awareness of what he has done. Apart from the question of its consistency with the tenets of retributivism this claim does not seem to be supported by the facts; conventional punishment is said by criminologists in fact to increase the criminal's resentment against society.

theory is often based on a psychological or sociological study of the effects of certain kinds of deprivation, cultural starvation, and general lack of education on the individual's outlook, and the hope is that these may be put right by re-education or psychological treatment in the period when there would otherwise be conventional punishment.

There are certain oddities about the reform theory if it is intended to be a justification for punishment. The first is that the processes of reform need not involve anything which is painful or unpleasant in the conventional sense. Some critics of the theory would rule it out on that ground alone. In reply, the advocates of the theory might say that in so far as the criminal is *compelled* to undergo reform the process can count as 'punishment' in the conventional sense; at least the criminal is deprived of his liberty and that is in itself unpleasant whatever else may happen to him during his period of enforced confinement. A different line of defence might be to concede that reform is not punishment in the conventional sense and to go on to point out that the reform theory is an attempt to replace punishment with a practice which has greater utilitarian justification. According to this line, punishment cannot be justified on utilitarian grounds and the utilitarian must therefore replace punishment with a practice which is justifiable.

There is a second respect in which the reform theory has consequences which are at variance with traditional ideas on punishment. The processes of reform may involve what might be called 'treatment'. It happens to be true that many advocates of the theory are influenced by psychological doctrines to the effect that the criminal is suffering from a disease of social maladjustment from which he should be cured. Hence, the processes of reform may include more than re-education in the conventional sense; they may involve what is nearer to brainwashing (and from a strictly utilitarian point of view there is everything to be said in favour of this if it is in fact effective). A merit of the retributive theory is to insist that persons as ends in themselves should be protected against undue exposure to the influence of moral 'doctors' no matter how socially effective their treatment may be.

The third respect in which the theory departs from traditional ideas is that it gives countenance to the suggestion that the criminal may legitimately be detained until he is reformed. But in some advanced cases of social disease the cure may take some time. And what of the incurables? Here again the retributive theory reminds us of the inhumanity of treating people simply as social units to be moulded into desirable patterns.

So far we have considered three oddities of the reform theory, but none of these, of course, invalidates it. The first point simply brings out the nature of the reform theory, and the second and third are hardly implications in the strict sense but merely probable consequences if the practice which the theory reflects is developed in a certain direction. The fatal defect of the theory is rather that it is inadequate as an account of the very many complex ways in which the sanctions of the criminal law are intended to affect society. The reform theory concentrates on only one kind of case, that of the person who has committed an offence or a number of offences and who might do so again. Moreover, it is plausible only for a certain range of cases in this class; those requiring treatment rather than conventional punishment. But it must be remembered that the criminal law is also intended for those, such as murderers, traitors, and embezzlers, for whom the possibility of a second offence is limited. Moreover, the criminal law serves also to deter ordinary citizens who might occasionally be tempted to commit offences. The inadequacy of the reform theory lies in its irrelevance to such important types of case.

It is clear, then, that no one of the accounts of punishment provides on its own an adequate justification of punishment. The retributive theory, which is often taken to be an expression of barbarism, in fact provides a safeguard against the inhumane sacrifice of the individual for the social good, which is the moral danger in the utilitarian theory. Bearing in mind this moral doctrine about the rights of the individual we can then incorporate elements from both the deterrent and the reform versions of utilitarian justification. Only by drawing from all three doctrines can we hope to reflect the wide range of cases to which the criminal law applies.

Nothing has so far been said about the justification of the concept of pardon. In fact there are no special problems about its justification, for pardon does not involve the infliction of unpleasant consequences on a person but the reverse. There is a problem of justification only for advocates of an extreme form of the retributive theory who hold that guilt is a necessary and a sufficient condition of punishment in the sense that whenever a person is guilty he ought to be punished. But, as we have seen, this theory does not commend itself to ordinary moral views even if we grant that ordinary views have a retributivist flavour to them.

5. CONCLUSION

A central theme of this book is that there is an important distinction to be drawn between what someone can do as a person and what he can do as a person-in-a-role. The value of the distinction has been shown in that it has enabled us to analyse the concepts of forgiveness and resentment, on the one hand, and punishment and pardon on the other. Having analysed the concepts we then tried to justify the practices they involved. Punishment in particular has traditionally provided a problem of justification. The argument of this chapter was to the effect that elements from the retributive theory must be combined with elements from the deterrent and reform versions of utilitarianism to cover the range of cases to which the law applies.

8

Moral Change and Moral Improvement

People sometimes speak of the present age as showing moral improvements in certain respects or as a whole over previous ages. More commonly, perhaps, they speak of the degeneracy of contemporary society as compared with that of the past. In this chapter I want to consider whether any precise sense can be attached to the ideas of the moral improvement and deterioration of a society, or whether we ought strictly to speak only of moral change. At the end of this investigation I shall try to offer some analysis of the term which features largely in popular discussions of this topic – 'the permissive society'.

1. MORAL PRACTICE AND MORAL STANDARDS

When people speak of the moral decline or improvement of a society, their evaluations may be referring to one or other of three different, although connected, phenomena: the actual conduct of the members of the society, their rules or standards, their institutions. Let us consider the logical connections between the phenomena.

To exemplify the first phenomenon let us assume that unpunctuality or absenteeism are widely thought to be morally bad; that there are generally accepted rules against them. Assuming such rules people might say that unpunctuality and absenteeism are on the increase, and conclude that in this respect the moral practice of a given society or sub-group of society is declining. Again, for some time there has been a common view that it is wrong to drive if one has had more than a certain amount to drink. People often paid lip-service to this rule but did not observe it in their practice. But now, it might be said, as a result of various advertising campaigns and legal threats, more people are coming to follow the rule in their practice, and hence in this respect society, or some sections of it, might be said to show a moral improvement.

If we approach the question of moral improvement or decline in this way it seems fairly straightforward. We simply assume a stable background of shared moral values and employ a social scientist to tell us what the statistics are in various sections of the community. Armed with sets of figures on absenteeism, alcoholism, drug addiction, church attendance, illegitimacy, prostitution, extra-mural class attendance, child cruelty cases, etc. we can make assured pronouncements on the moral position of our own society as compared with that of a bygone age. This simple view of things is all that most people have in mind when they deplore the morality of contemporary society, and the main talking points raised by this approach concern only the accuracy and relevance of the statistical figures.

But quite apart from the matter of statistical accuracy there is a criticism which can be made of this simple view of moral improvement or decline. The difficulty is that there is a much closer connection between moral rules and moral practice than seems to be implied by the first view. It is true that philosophers differ over the connection between moral rules and moral practice: some assert that the connection is one of logical necessity, in the sense that if a person claims to adhere to a moral rule then he logically must keep to it in practice; if he does not keep to it then it may be said that he does not fully accept it as a moral rule. Strict forms of this view – sometimes called 'prescriptivity' – have their own difficulties, mainly concerned with making sense of the phenomenon known as 'weakness of will'. For, if a person does not in fact keep to a moral rule, then, according to the prescriptivist view, he does not in the full sense accept the rule as binding on him. It cannot therefore be true that his will is weak since he does not hold the rule against which his will might be judged weak. But even if some philosophers reject the thesis of prescriptivity, that there is a tight *logical* connection between professing a moral rule and keeping to it in practice, most philosophers would agree that there must be some kind of close connection between moral rule and moral practice, however it is to be analysed. It is not simply a happy contingency that people often guide their conduct by their professed moral rules, but rather that that is the purpose of moral rules.

Now if we accept that, however it is to be analysed, there is some sort of close connection between moral rules and moral practice, then the view of moral improvement or decline so far presented must at least be oversimplified. For, if it is the case that large sections of society do not in their practice reflect the guidance of a certain moral rule, we can ask whether that rule is in fact still a moral rule accepted by the society. In other words, there are difficulties attached to the claim that a society has declined with respect to a given rule in that, if 'declined' implies that they no longer keep to the rule, then that very implication might lead us to say that the rule is no longer a rule of the society, and hence that the society cannot be said to be declining in respect of it. One and the same factor – namely, a set of figures or other evidence about a moral change – can therefore feature in two different ways in an argument about moral decline. In terms of the original simple view the moral change is used as an empirical premise which, together with the assumption of the operation of a rule, gives grounds for inferring a moral decline. In terms of the second more sophisticated view the moral change is used as a premise in a necessary or *a priori* argument to the effect that such and such a moral rule logically cannot be held by a society since the practice of the society does not seem to be guided by it.

In a somewhat similar way the argument could be used to cast doubt on a simple view of the nature of moral improvement. If a society pays lip-service to a moral rule, such as that one ought not to drive after having drunk alcohol, but does not adhere to it in practice, then it may be doubted whether the members of the society hold it as a moral rule at all. If later the drivers in the society tend to keep to such a rule we can say that they now hold it as a rule. But this cannot count as improved practice with respect to their adherence to the rule since previously they did not accept the rule as such.

The complications which have been raised by means of the *a priori* argument do not invalidate the first simple view of moral improvement or decline but rather focus attention on the second of the phenomena which, as I claimed at the outset, is sometimes the subject of such discussions – the rules or standards of the members of a society. The question now becomes that of deciding whether it makes sense to speak of moral improvement or decline in rules or standards themselves. Indeed, the common phrase 'declining moral standards' is ambiguous: it may refer to declining standards of adherence to a given rule or set of rules, or it may refer to decline in the rules themselves, in the sense that where previously there was an operative rule there is now none, or a morally inferior one. The tendency of the *a priori* argument, then, is not to invalidate the first account of moral decline and improvement but to bring out that the answer to the first question of whether actual practice can be said to improve or decline with respect to rules, depends logically on the answer to the second question, whether the rules themselves can be assessed morally.

I have so far posed the problem in terms of the moral improvement or decline of a *society*, but precisely the same issues arise over the moral improvement or decline of an *individual*. If we take the first phenomenon – the actual moral practice of the person – then we may say that it has improved or declined with respect to some rule or rules. But since there is more than a contingent connection between practice and rules, since, in other words, the nature of a man's practice has a logical relevance to the question of what rules he accepts anyway, we are forced to give a logical priority to the question of whether it makes sense to speak of rules themselves as improving or declining morally. Before we turn to that problem, however, let us consider the third phenomenon sometimes mentioned in discussions of the improvement or decline of social morality.

It is sometimes said that a given society is improving or declining morally with respect to the social, political, or legal institutions which it possesses or introduces. For example, it might be said that a society with democratic political institutions is morally better as a society, irrespective of the actual practice of its politicians, than one without such institutions; and that if the society abandons its democratic institutions it has declined morally no matter whether the actual practice of the politicians is adversely affected or not. Again, it might be said that a society with secret policemen (or even, more controversially, armed policemen) is morally inferior to one whose policemen are uniformed and unarmed. Clearly, examples of this kind can be multiplied. They are all, however, logically dependent for their force on the decision whether or not we can say that moral rules themselves can be better or worse. For in judging that a certain social institution is a morally good thing – say, that it is a morally good thing to have a welfare state – we are presupposing that certain moral standards can be said to be an improvement on others. We judge institutions no less than practice in terms of standards.

It seems, then, that of the three phenomena often mentioned in ordinary discussions of the moral improvement or decline of society – changes in practice, in rules and in institutions – it is the second which has the logical priority. Let us therefore turn to the question of whether it makes sense to say that rules or standards can improve or decline morally.

2. SUBJECTIVITY, OBJECTIVITY, RELATIVITY, ABSOLUTE VALUES

It might be said that if rules or standards are 'subjective' then we cannot speak of their improvement or decline but only of their change, whereas if they are 'objective' we can speak of improvement or decline. In this kind of context 'subjective' means 'reflecting only the opinions or feelings or tastes of individuals'. Thus, the question of whether or not a certain dish is 'good' is in the end subjective or dependent on the gustatory likings of people. Of course, the dish may or may not be good in the different sense that it is or is not a good specimen of *boeuf stroganoff*, *zabaglione* or the like, and on such questions the advice of the experienced is required and a decision may be reached which, within a certain framework, is not subjective, or not entirely so. But the basic question of whether a good example of the dish is itself good remains a question the answer to which is a matter of taste or is subjective. Now if moral rules are ultimately subjective in this sense then it does not seem that one can speak in any fundamental sense of moral improvement or decline. Tastes change, feelings vary, and fashions alter, but it is not clear that one fashion, feeling, or taste can legitimately be called better than another.

Let us now consider whether we are enabled to speak of moral rules and standards as improving or declining if we regard them as objective. The word 'objective' is ambiguous. It might mean, in the first place, 'impartial' or 'influenced only by those considerations which ought to be influential'. It is in this sense that examiners or judges might be said to be objective, for they judge according to rules which they apply in the same way to all. But this sense of 'objective', although it is a legitimate one, does not enable us to speak of rules themselves as objective for, while it clearly presupposes the existence of rules, it does not imply anything at all about their status as rules. A judge can *apply* a rule objectively or impartially although the rule itself owes its existence to the arbitrary whim of a tyrant. If we said of the rules of this tyrant that they showed a moral decline from those of the previous ruler we would be using some standard other than that of objectivity-as-impartiality, for the judge could be equally impartial about the application of either set of rules.*

* cf. the third sense of 'impersonal', pp. 135–6.

There is a second sense of 'objective' which seems nearer to our purposes. In the second sense, 'objective' means 'capable of truth or falsity'. Statements purporting to express facts are objective in this sense; they can be true or false or, in philosophical jargon, they can have a truth-value. If we say that morality is objective in this sense we shall be implying that what is logically prior is not a rule but a moral judgement, such as 'Truth-telling is right'. This judgement will be objective in the sense that it will be true or false. And the rule 'One ought to tell the truth' will be objective in the sense that it expresses the truth of the prior judgemcnt. Does it follow that if morality can be said to be objective in this sense (and it is a view which many philosophers would reject) that we can speak of moral improvement or decline?

The answer to this question depends in the first instance on the answer to the question of what it is that makes moral judgements true or false (if they are such). Suppose we say that 'right' means 'what gives pleasure'. Such a definition of a basic moral term will enable us to regard moral judgements as true or false, for a judgement such as 'Showing gratitude is right' will mean 'Showing gratitude causes pleasure' and this judgement will be either true or false. Similarly, 'Spreading malicious gossip is wrong' will mean 'It causes pain or displeasure to spread malicious gossip', and this will be either true or false. Thus a definition of morality in terms of its conduciveness to the creation of pleasure enables us to have objectivity in the relevant sense, but it does not enable us to speak of moral improvement or decline. The reason is that pleasure is itself something subjective, depending as it does on the individual responses of people to their situations. Hence, if we are to speak of moral improvement or decline we shall need to say that moral judgements are objective in the sense that they are capable of being true or false, and also that what makes them true or false is something other than subjective responses. To stipulate in this way is not, of course, to rule out all so-called 'naturalistic' definitions of morality, but only those which define morality in terms of the feelings of arbitrary choices or other subjective responses of the agent. It would be quite possible for a philosopher to define morality as a means of harmonizing interests on a co-operative basis, or of furthering evolution, and (whatever the objections to these definitions on other grounds) we should have objectivity in the sense which would enable us in theory to speak of the moral progress or regression of a society. Thus only certain kinds of 'naturalistic' definition of morality prevent us from speaking of moral improvement or decline. Let us now consider whether the possibility of speaking of moral improvement or decline requires, in addition to the presumption of one kind of objectivity, the presumption of the absolute nature of morality, or whether a relativistic account still enables us to think of morality as being better or worse.

An objectivist position which was also absolutist would be to the effect that moral judgements are true or false in virtue of some non-natural fact, some absolute value or Platonic Form which never changes. The best statement of the absolutist position on morality is in fact that of Plato. Plato held that behind the flux of everyday experience there are certain patterns or blueprints or ideals of all that exists. For each bed there is one Form of the Bed and for each just action there is one Form of Justice. This doctrine of Forms culminates in the Form of the Good which Plato compares to the sun. The Form of the Good, like the sun, sustains the actual world, lightens it and is the ultimate source of health and good. Moral judgements will be true or false in so far as they do or do not fully express the Form of the Good.* All the possible versions of absolutism in morality are simply versions of Platonism.

To this thesis – that an absolute view of morality is Platonic in nature – it might be objected that there are religious versions of absolutism which make the ideal of God's will or command central. For example, it might be said that a version of absolutism could assert that actions are right or wrong absolutely if God commands them. But this thesis is

* Plato, *Republic*, Book VI.

ambiguous. It may mean, in the first place, that actions are right because God commands them, or it may mean, in the second place, that God always commands right actions. The first interpretation makes the command logically prior to the rightness – the command creates the rightness, as it were – whereas the second interpretation makes the rightness prior to the command – it is because the action is absolutely right that God commands it.

The first interpretation is open to an objection. The objection is simply that it always makes sense to question a commander. If a commanding officer in the army orders a recruit to do something it always makes sense (although it may be disastrously imprudent) for the recruit to question his commander. And this holds for any commander, even for God. One can always say, 'God has commanded this, but is it really right?' Some might argue that this is what Abraham ought to have said to himself when God commanded him to sacrifice Isaac. It would certainly have made sense for him to have questioned the rightness of the command. But if it even makes sense to question the rightness of such a command then actions cannot be right simply because God commands them. Those who think that actions can be right simply because God commands them are probably implicitly assuming the judgement 'God is righteous'; but this judgement presupposes that rightness or righteousness is defined or in some other way exists independently of God's commands. And to assume this is to assume the second interpretation of the claim that God always commands what is right. But the second interpretation is once again a version of the Platonic idea that there exist absolute values or Forms. It is to assume that rightness exists independently of anyone's will, even God's, and that moral judgements are true or false to the extent that they embody these Forms. Many religious accounts of morality are simply a confused combination of both interpretations; they assume that because God is a guide to what is right he somehow also creates rightness.

Be that as it may, it is clear that if we have an absolutist view of morality it makes sense to speak of moral improvement or decline. In so far as the moral rules of a given community at a given historical time approximate to these absolute values, to that extent the community will show an improvement over another community, or the same community at a different time, where the rules less fully express the absolute values. Such a conception clearly gives a sense to the idea of moral improvement or decline. There are considerable difficulties in the view, but, before we raise these, let us consider whether a relativist account of morality enables us to speak of moral improvement or decline.

A relativist account of morality can be objectivist. The relativist will hold that moral values reflect the economic and general cultural outlook of a society. To say this, however, is not to say that they depend on the arbitrary taste or choices or feelings of any one individual or group. The values may be embodied in a total way of life, but to be relative to a way of life is by no means to be variable, unstable, or wayward. Moral judgements on a relativist account can be true or false in so far as they reflect the values of the community.

Now the values of a community can be objectified in its characteristic way of life, so that it is possible, without too much bad metaphysics, to think of the 'spirit' of a people. If it is possible to think in this way then it is possible to say that the rules or moral practice of a given period more fully embody that 'spirit' than the rules or practice of another historical period. In fact, historians – and not just philosophers of history – do speak in this way of golden ages in the life of a community, of the 'flowering of the genius' of a people, of the 'decline' of an empire or the like. Such judgements of periods in the histories of Greece, Rome, Venice, Florence, France, and so on are common. And they seem to be judgements of moral (among other) improvements or declines, but they do not on the surface seem to presuppose an absolutist view of improvement or decline. They seem rather to presuppose the possibility of discovering in some rough and ready way what is the particular genius or 'spirit' of a culture and using that as the norm to judge improvement or decline. It may therefore be contended that relativism is not incompatible with the possibility of making judgements of moral improvement or decline.

There are two difficulties in this attempt to combine judgements of improvement or decline with relativism. The first is that the view is by its very nature severely limited in its scope. It is incompatible with judgements of the moral rules of one community against those of another or of one type of morality against another. If we assume relativism we can speak only of *moralities* and not of *morality*. And whereas the concept of 'a morality' allows us 'internal' or 'vertical' evaluation it does not allow us 'external' or 'horizontal' evaluation of one morality against another. For the latter possibility presupposes the concept of 'morality' as distinct from 'moralities'. If we are to be able to speak in the full sense of moral improvement or decline we therefore require to assume some absolute view of morality.

The second difficulty is even more radical, for it seems that the view does not even allow us 'vertical' evaluation of a morality without presupposing an absolutist view. Suppose we say that, at a given period in its history, a people revealed more of its true nature, seemed more highly developed, had finer art, nobler institutions and the like. If we say this, a question can be asked as to why such developments are considered better than others where there is little art but a great deal of bread, circuses, absentee landlordism, oppression of the poor, etc. The answer in the end can only be a repetition of the moral judgement that the former developments are better. And in this context 'better' must be analysed in terms of some absolutist view of morality.

It seems, then, that there is no middle way between subjectivism, which does not allow us to speak of moral improvement or decline, and absolutism, which does. It would be inappropriate in a work on social ethics to argue for an absolutist view of morality, so we shall simply assert that in the type of moral view with which we are here concerned there is presupposed the truth of the view that what is supremely important is the individual and his self-development, or, in Kant's formula, that we ought to respect human nature whether in our own person or in that of another. It is in terms of this view that people in liberal democratic culture make judgements that society is improving or declining morally.

Even supposing we assume the truth of this 'respect for persons' view of morality, however, there is still considerable difficulty about deciding whether or not any given moral change is in fact an improvement or a decline with respect to the standard. The difficulty is due mainly to the fact that there is some difference of opinion from one period to another as to what in fact constitutes human self-development. As an illustration of this let us consider the analysis of a change in morality which has come to be known as a change in the direction of 'permissiveness'.

3. PERMISSIVE MORALITY

In a discussion of 'permissive morality' it may be helpful to raise questions such as the following: What phenomena do people have in mind when they speak of 'permissive morality'? What do these phenomena have in common? Can we say (assuming that 'respect for the individual' is the ultimate test of the acceptability of a moral position) that a move in a permissive direction constitutes an improvement or a decline in moral conduct? Is contemporary morality adequately characterized in terms of 'permissiveness'?

In answering the question of what people have in mind when they speak of present-day permissive morality we might plausibly refer to a wide range of phenomena.* For example, they are often thinking of the whole area of sexual morality. Thus, they might be holding that sexual relationships are now regarded as legitimate outside marriage. Again, they might have in mind that sexual topics are freely discussed in ordinary conversation or in the newspapers or on radio and television. Again, they might be claiming that novels,

* cf. C. H. and Winifred M. Whiteley, *The Permissive Morality* (London: Methuen, 1964).

plays, films, and other art forms treat the sexual side to human conduct in great detail. It cannot be denied that such factors are often in the forefront of people's minds when they use the term 'permissive', nor that, as a matter of fact, there is a good deal of substance in the claims.

But the change in sexual morality over the last thirty years is only one of the factors which might be mentioned as making up permissive morality. We find, for example, that people speak of the 'permissive family'. What seems to be involved in this conception is a change in attitudes to discipline, to the rights of the children and the parents, to the opinions of children and so on. Children are encouraged to have and to express views on what they will do, say, or wear. These views may differ from those of their parents and it is not thought that parental views should necessarily be enforced. Indeed, questions of enforcement apart, it is not felt by either parents or children that parental views on many matters are necessarily the correct ones or the best ones. Again, children are provided with large sums of pocket-money and see this as a right rather than as a favour. Many other specific features of the present family situation could be mentioned, but perhaps enough have been mentioned to identify a contemporary trend in the development of the family.

Another area to which the term 'permissive' is sometimes applied is that of education. The present-day classroom is sometimes depicted as an arena of permissiveness. For example, in many primary schools the desks are not arranged in neat rows but in groups distributed round the room; pupils are allowed to wander about the classroom to look at the nature table, the library, etc. They are not forbidden to talk to each other and they may be on terms of easy familiarity with their teacher. The rote-learning of multiplication tables may be replaced by the measuring of desks and classroom floor with rods, and the listing of the principal products of Denmark may give way to some sort of 'project' aimed at providing a total picture of the life of the country

There are many other areas of modern life which are said

to be moving in a permissive direction. We find, for example, that people sometimes speak of 'permissive prisons', of the welfare state as permissive, of churches as becoming permissive and so on. I believe that it would be generally agreed that these changes are taking place and that people, however they might differ on details, would agree roughly on the nature of the changes. The question which arises is that of whether there is any general description which would sum up the changes.

This question is harder than it might seem, because a general description of the changes (as distinct from detailed descriptions of specific changes) may tend to beg questions about their evaluation – whether they represent an improvement or a decline – and the evaluation of the changes is a topic on which there can be much disagreement. For example, if we describe the changes as 'giving the individual a much greater degree of freedom' we may be implying a favourable evaluation, in that the word 'freedom' has strongly favourable associations. On the other hand, if we describe the changes as 'all part of the general social slackness' we shall have implied an unfavourable evaluation. A first step in clarifying the issues on this matter may be taken if we make use of a distinction introduced in the first part of the chapter – the distinction between the weak observance of a non-permissive standard and the assertion of a new permissive standard.

In any large-scale social change two different phenomena may be found operative. The first is that many people will continue to pay lip-service to rules or values they have been brought up to honour, but will in their actual conduct tend to disregard these rules. The rules or values will be mentioned on Sundays, as it were, but not during the week. Whether or not they can be said to accept the rules or values in any meaningful sense is (as we have seen) a complex question the answer to which will depend on such matters as the extent to which they feel guilt if criticized by defenders of traditional values, the extent to which they teach the old values to their children and so on. Clearly there will come a time when their judgements will simply be a form of words having no bearing on their conduct, and we shall need to say, even if we do not accept the full-blooded thesis of prescriptivity, that their words belie their actions and that it is in their actions that we discover their true moral beliefs.

The second element in a large-scale social change will be contributed by the bolder souls. At a much earlier stage in the change they will explicitly be asserting the new outlook, formulating its rules and so on. They will be providing the reflective commentary on the changes and the 'experiments in living' which may go with them.

Now if we are thinking mainly of the first element in social change it is understandable if we characterize it in some pejorative way. People who claim to accept certain rules may fairly be described as 'slack' if their conduct does not reveal this. But where we are thinking of the second element we can less easily use such descriptions-cum-evaluations; whatever is wrong with an *avant-garde* group it can hardly be described as 'slack'. It is indeed a feature of such groups that they are often highly critical of the rest of society, describing it perhaps as hypocritical or as insincere. For example, D. H. Lawrence's criticism of the sexual morality of his society consists of a conscious assertion of certain values of sincerity in human relations as against what he took to be the hypocrisy and inhumanity prevailing in the industrial society of his times.

Let us assume that we are dealing with the second of the two elements involved in the change to a permissive morality, and let us describe this change as one to a state of affairs where the onus is on the individual to make up his own mind on what he is going to do from a state of affairs in which his conduct was to a great extent decided for him by rules of one kind or another. Assuming that the supreme standard of morality is respect for the individual, can we say that the change to a permissive morality, as we have specified it in detail and in general, represents a decline or an advance in morality?

A first reply to this might be that in so far as permissive

morality gives to the individual great scope for his own reflection and decision it represents *ipso facto* a moral advance over a system of morality in which there is blind obedience to rules. Strong support for this assessment can be found in the writings of Existentialists, who see in the exercise of authentic or reflective choice man's only good, and in the unreflective conformity to accepted customs and rules *mauvaise foi*, his only evil. Similar sentiments are expressed by Socrates when he says that the unexamined life is not worth living. Many people, of course, would not go so far as the Existentialists or as Socrates in their assessment of the importance of individual choice and reflective consideration of the basis of one's moral beliefs, but most people would probably agree that the reflective life expresses the true nature of a person, is more characteristically human, than the unreflective. Hence, if it is the case that permissive morality puts greater stress on the individual's own thoughts and decisions than a more rule-bound form of morality then it does create a framework in which moral advance is possible.

On the other hand, it also creates a situation in which an individual may simply go along with the desires of his social group without ever raising with himself the question of whether he himself really wants to follow the ends which others are pursuing, let alone whether he thinks that in his situation he really and truly ought to follow them. Hence, if permissive morality creates the possibility of an increase in reflective living it also creates the possibility of a new conformity – not the hidebound conformity produced by rules, but the (equally mindless) conformity produced by following the fashionable urgings of in-groups; one's very desires, let alone one's obligations, are nowadays identified for one by 'briefing' in fashionable Sunday newspapers.

Moreover, a society in which rules are at a low premium creates other moral difficulties. Let us consider a fairly trivial example of the difficulties as they arise within the sphere of etiquette.

It would be agreed that one manifestation of the permissive outlook has been a great slackening of the rules of social etiquette. There is no uniformity of dress for special occasions, and there are no agreed methods of introduction. Now the advocates of this breakdown of rules of etiquette will claim that fresh air has been let in to clear the stuffy conventions of former years. But, as against this, it must be remembered that the point of having rules of etiquette is to guide people on how they ought to behave; if there are no rules then there is the possibility of greater embarrassment. For example, if there is a rule that one ought to shake hands on being introduced then introductions are free of one problem. But in our present social situation there is no agreed procedure and the result is a certain awkwardness and lack of grace in a social situation which is in any case not an easy one. In short, where there are few rules of etiquette the possibility of awkwardness and embarrassment, and hence of hurt feelings of all kinds, is greatly increased. And this, although a trivial matter, illustrates a disadvantage of a permissive outlook.

Let us now consider a more serious example of the same point. Consider the situation of a young couple embarking on marriage. In former days there were clear rules dividing the labour of husband and wife within the home, and as between the outside world and the home. Nowadays, however, all this has been altered and young couples must decide for themselves what the rights and duties of their own marriage will be, what the role of each will be. The moral opportunities of this are obvious; there is a chance for each to shape the marriage in a way which will realize individual potentialities. Just as often, however, this opportunity is a source of unhappiness and strife within the marriage. It may be difficult for each partner to accept the role of the other, or one may expect more than the other. In other words, the freedom brings with it tensions and anxieties which would have been absent in a marriage-situation which had firmly defined rights and duties.

Somewhat similar considerations apply to some of the other areas of conduct in which permissiveness is said to prevail. In sexual morality, for example, a permissive attitude may simply mean following mindlessly the behaviour patterns of some dominant social group. If this is what it means it clearly does not on any showing constitute a moral advance. But if it means a commitment to serious consideration of what is appropriate in sexual behaviour in a given situation then this may well constitute a moral advance on an unswerving adherence to rules. But the price of this freedom may be a great increase in tensions and anxieties of all kinds. Thus, if a person is convinced that sexual intercourse outside marriage is wrong, then, however difficult he may find this rule to observe, he is at least clear what he *ought* to do. A permissive morality, on the other hand, brings with it many doubts as to what, morally speaking, one ought to or may do; in one way this can be liberating, but in another it can be crippling in the anxieties it carries with it.

In sum, then, we may say that if we assume that respect for the individual and his essential features as a person is the basic principle of morality we may also claim, although with many reservations, that what is called permissive morality constitutes a moral advance over a rule-bound view of morality in that it provides greater freedom for people to express their own individuality and to develop their potentialities. In saying this, however, we have not settled another question which may be raised over the present state of contemporary moral outlooks, namely whether these outlooks are exhaustively characterized by the term 'permissive'. Let us, in conclusion, briefly consider this question.

It seems to me that the term 'permissive' conveys no suggestion of two of the most important moral changes of the century: the change in the place of women in our society, and the change in attitudes to war. It would be inappropriate for a philosopher to try to document in his amateurish way the radical transformation which has taken place in the lot and status of women over the last sixty years. Fortunately, this is not necessary for our purposes, because as soon as the point is mentioned its truth is obvious in a general way. What is not prima facie obvious, but is none the less self-evident after reflection, is that this change in the female lot constitutes a *moral* advance. For, if it is agreed for purposes of argument that our basic moral principle is that of respect for the individual, then clearly the transformation of the status of half the human race – the female half – can fairly be considered an advance in morality. This claim seems to me so obviously justified that it is strange that it is so often left out of account in a reckoning of the present-day moral outlook.

A second factor often left out of account in such reckonings is the change in the attitudes to war which has taken place over the last sixty years. To say this is not to claim that there are fewer wars or that they are less cruel. It is to say, however, that Western governments at least no longer receive uncritical support from public opinion for aggressive foreign policies. As examples of this claim we might cite the division in British public opinion at the time of Suez and the division of U.S. public opinion over policies in Vietnam. The fact that public debate on such matters has become possible at all is a sign of an important moral advance on this matter over previous centuries. Any characterization of contemporary moral trends ought to mention this as well as the move or drift in a permissive direction.

4. CONCLUSION

In this chapter I have been considering whether any sense can be given to the idea of moral advance or decline. There are various theoretical difficulties in the way of such an idea, and it seems that we must assume not only that morality is objective but also that there is an absolute moral principle before we can give any firm sense to the idea. Assuming that the principle of respect for the individual and his development is the principle which is likely to be regarded by liberal-democrats as absolute (if any is), I examined some of the most obvious features of present-day Western morality to see whether it could be said to be an advance or a retreat on the morality of sixty years ago. The conclusion which emerged is that present-day social trends – which may be described as 'permissive', provided we also remember the factors of the place of women and attitudes to war – may be said to offer at least the possibility of a moral advance, although such a possibility carries with it the counter possibility of new kinds of degeneracy. Perhaps on this note of qualified optimism we should bring the discussion to an end, in case further arguments suggest themselves to spoil the hopeful picture.

Postscript

In this book I have tried to link moral, social, and political ideas in a single philosophical theory. The religious and metaphysical implications of the theory are not here developed in any detail, but in Chapters 1 and 8 I have tried to indicate their general nature. The arguments by which I have arrived at my position are doubtless open to many criticisms, but I should like to end by admitting to an unease about another aspect of the theory: we can call it the ideological aspect.

I have attempted to restate the viewpoint of liberalism in a way which incorporates the insights of mid-twentieth-century socialism of the 'welfare state' variety. This has involved stretching the principles of *laissez-faire* liberalism to accommodate the ideas that the Government, in its welfare legislation and other matters, has a paternalistic function, and that it ought not to leave morality behind in foreign policy. I have argued that these moral ideas develop out of liberalism and are reasonable extensions of liberal doctrines. It may be, however, that liberalism cannot accommodate such ideas and that they require a totally different theoretical framework. If this should be the case then it will not be possible to synthesize the principles of *laissez-faire* liberalism and mid-twentieth-century socialism, and my attempt will turn out to be radically ambiguous. I do not know whether this is so or not.

BIBLIOGRAPHY

This book has been influenced by many contemporary writings on social, political, and moral philosophy, but I am most conscious of the influence of the following:

Ernest Barker, *Principles of Social and Political Theory* (Oxford: Clarendon Press, 1951).

S. I. Benn and R. S. Peters, *Social Principles and the Democratic State* (London: Allen & Unwin, 1959).

Dorothy Emmet, *Rules, Roles and Relations* (London: Macmillan, 1966).

H. L. A. Hart, *The Concept of Law* (Oxford: Clarendon Press, 1960).

D. D. Raphael (editor), *Political Theory and the Rights of Man* (London: Macmillan, 1967).

Other Recommended Books

H. B. Acton (editor), *The Philosophy of Punishment* (London: Macmillan, 1969).

Dorothy Emmet, *Function, Purpose and Powers* (London: Macmillan, 1958).

H. L. A. Hart, *Law, Liberty and Morality* (London: O.U.P., 1963).

H. L. A. Hart, *Punishment and Responsibility* (Oxford: Clarendon Press, 1968).

Anthony Quinton (editor), *Political Philosophy* (London: O.U.P., 1967).

J. P. Plamenatz, *Consent, Freedom and Political Obligation* (Oxford: Clarendon Press, 1938).

J. P. Plamenatz, *Man and Society* (London: Longmans, 1963).

D. D. Raphael, *Problems of Political Philosophy* (London: Pall Mall Press and Macmillan, 1970).

Index